Early Praise for *Build, Tune, Explore with OpenWebRX+*

I wish I'd had a book like this when I got started with software-defined radio. Richard has created a resource that will let most people pick up an SDR and make real progress quickly.

➤ **Mark, VK2MOH**

The book is logical and easy to follow and will certainly let a beginner to networked SDR radio get up and running in very short order. Even more advanced radio hobbyists will benefit from Richard's insights especially with regards to making an SDR accessible to the online world via the Internet.

➤ **Gerald Hynes, VE4ACE**

I found this book to be a good reference for those interested in starting out in OpenWebRx+. The step-by-step instructions for setting up a Raspberry Pi were detailed and complete and the reference links were particularly useful.

➤ **Andrew Hook, VK2EAH**

It was a pleasure to read. All the information that you need to set up an SDR station is well presented. A book to keep.

➤ **Jan Van Ekris, VK2FEB**

Build, Tune, Explore with OpenWebRX+

Web-Connected Software-Defined Radio Made Simple

Richard Murnane, VK2SKY

The Pragmatic Bookshelf

Dallas, Texas

Pragmatic Bookshelf

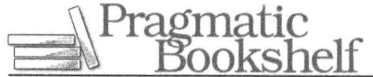

See our complete catalog of hands-on, practical,
and Pragmatic content for software developers:
https://pragprog.com

Sales, volume licensing, and support:
support@pragprog.com

Derivative works, AI training and testing,
international translations, and other rights:
rights@pragprog.com

The team that produced this book includes:

Publisher:	Dave Thomas
COO:	Janet Furlow
Executive Editor:	Susannah Davidson
Development Editor:	Kelly Lee
Copy Editor:	Karen Galle

ISBN-13: 979-8-88865-178-0
Book version: P1.0—December, 2025

To my wife, Miki, who has put up with this
ラジオ 親父 (rajio oyaji) above and beyond the
call of duty, and to our beloved son, Sean.

Contents

Foreword

You might not know it, but you're about to discover the joy of radio.

Radios of all sorts are constantly broadcasting signals. Signals with audio. Signals with images. Signals with data. You're familiar with some of these signals—after all, you've probably listened to music in the car, watched television, or used Wi-Fi before. But there's so much more out there to discover. Signals are everywhere and they contain all sorts of interesting things.

This book is the first step in unlocking it all. Richard starts you out with just listening to the radio, the FM variety with all the hits, but quickly dives into all sorts of signals. He shows you how to listen to air traffic control, look at pictures transmitted from the ISS, and decode the data that aircraft broadcast containing their location, heading, and speed.

A software-defined radio with an antenna, OpenWebRX+, and a small computer like a Raspberry Pi are all you need to get started. But radio is a big hobby and, frankly, it can be a bit fiddly. It can be hard to know where to start and you can get stuck trying to troubleshoot things you don't fully understand.

This book gives you a great starting point, a path to follow, and help with all the fiddly bits. Richard explains enough of the theory so that you understand what you're doing without overwhelming you. It's everything you need to get started with software-defined radio—well, besides the hardware itself—and is way better than watching random YouTube videos until you figure it out.

Richard's approach to introducing you to radio mimics my own journey. I first discovered software-defined radio from a collection of websites that were running software similar to OpenWebRX+. Unnamed nerds of the highest order had hooked software-defined radios to the Internet so that this nerd could play with them.

It was amazing! I could see signals in the waterfall display. I could tune and modify and control every aspect of decoding, demodulating, and listening to those signals. And there were so many signals out there hiding in invisible light just waiting to be discovered. It was like finding a new world.

I fiddled without understanding and began to learn. I bought my own software-defined radio. I bought and built antennas. I listened to FM radio (with all the hits). I listened to air traffic control. I looked at images from the ISS. I decoded aircraft transponder signals.

You, too, will do these things. And more.

I eventually grew beyond my software-defined radio. I wanted to start inter-acting with these signals. And so, I went and took the tests and got my ama-teur radio operator's license. My friends now call me W8GUY (or Whiskey-Eight-Golf-Uniform-Yankee if they're extra cool) and I talk to people all over the world.

And it all started because of a nerd, their website, and a software-defined radio. I wish I had had this book when I started the journey, but I'm glad I took it regardless. I hope you enjoy your journey into radio as well, and I hope to someday hear you on the air.

Guy Royse, Nerd, W8GUY, August 2025

Acknowledgments

According to an old proverb, it takes a village to raise a child. The same can be said of writing and publishing a book, so let me acknowledge my fellow villagers.

First, the dedicated radio amateurs and software developers whose work led to the creation of this book:

- András Retzler HA7ILM,[1] who developed the original OpenWebRX[2] as a PhD project;
- Jakob Ketterl DD5JFK, who forked the GitHub repo and gave the project new life;[3]
- Marat Fayzullin KC1TXE, who further forked the repo to create OpenWebRX+,[4] a version with many valuable enhancements;
- and John Seamons ZL4VO/KF6VO, creator of the popular KiwiSDR,[5] another web-connected SDR based on OpenWebRX. My work configuring a Kiwi for my radio club, the Manly-Warringah Radio Society,[6] led to my book proposal to The Pragmatic Programmers, speaking of whom …

My developmental editor, Kelly Lee, patiently and expertly guided my transition from an aspiring author to a perspiring one, and hopefully now, an inspiring one as well! Also, to the Pragmatic Programmers Art Department for their delightful cover design (Eddy the Gerbil says "73!"—if you know, you know). My copy editor, Karen Galle, adapted my working-from-home Australian King's English to American ("zees" abound where once there were "esses") and polished many rough edges along the way. And Dave Thomas provided much-needed feedback and kind words of encouragement.

1. https://blog.sdr.hu/2019/12/29/openwebrx-eol.html
2. https://github.com/ha7ilm/openwebrx
3. https://www.openwebrx.de/
4. https://fms.komkon.org/OWRX/
5. http://kiwisdr.com/
6. https://www.mwrs.org.au

A number of my fellow amateur radio operators conducted technical reviews of the pre-beta book and provided invaluable feedback. In call sign order:

- Gerald Hynes VE4ACE
- Craig Pattison VK2BTQ
- Andrew Hood VK2EAH
- Jan van Ekris VK2FEB
- John Vetters VK2JV
- Mark Halliwell VK2MOH
- Rajesh Doolabh VK2VKP
- Guy Royse W8GUY,[7] who also kindly wrote the book's Foreword.

Craig and John were particularly helpful in relentlessly prodding me for progress reports during our daily coffee runs.

Finally, a special shout-out to Volodymyr "Wlad" Gurtovy US7IGN,[8] author of *War Diaries: A Radio Amateur in Kyiv.*[9] I learned that if Wlad could successfully write a series of books in the middle of an active war zone, I could probably manage to write one from the comfort of my home in northern Sydney, Australia.

Thank you, one and all!

---

7. https://guy.dev/
8. https://www.us7ign.com/
9. https://www.amazon.com/dp/B0BLZM8H9P

Preface

Hi, I'm Richard! My radio friends know me as VK2SKY. I've been a radio enthusiast since my childhood in New York, when "Santa Claus" left my brother and me a pair of walkie-talkies one Christmas morning. Little did he know what he was starting. Or maybe he did ... seven years later, I found myself studying electronics at the University of Limerick, in Ireland, and spent many years in the electronics and software industry. Many years ago, I migrated to Sydney, Australia. Radio can take you a long way!

Fast forward to the present, and radio is almost unrecognizable, yet familiar. We still have walkie-talkies—HTs or "handy-talkies" we call them these days—and almost everyone listens to voices and music on the radio. But applications of this magical technology have advanced far beyond that, and radio is now a part of everyone's lives, whether they realize it or not.

In this book, we'll go on a little journey together, into the realm of "web-connected software-defined radios," or webSDRs for short. Using little more than a Raspberry Pi microcomputer, a USB radio receiver dongle, and a basic antenna, we can explore the far reaches of the radio spectrum and discover its countless secrets. And, if we like, we can share those secrets with the world.

Only a few years ago, this kind of exploration required expensive dedicated hardware, and as a ham radio operator, I've gone through my fair share of radios over the years! But with a web-connected SDR, we can do it all (receiving, if not transmitting) with an inexpensive kit small enough to fit in a shoe box.

Today, you can choose your own radio adventure: listen to air traffic, track nearby flights, ships, and vehicles on a map, receive pictures from the International Space Station, discover digital signals of all kinds, and much more. And if the signals you want to chase are out of range of your webSDR, there are hundreds more around the world, as close as your web browser.

Who Is This Book For?

You don't need to be an electronics or radio expert to play here, nor a computer whiz; if you know your way around a Linux terminal, that's great, but if not, follow the examples and you'll do fine. Along the way, we'll have short, simple projects to build understanding and skills.

My aim here is to get you up and running OpenWebRX+ with minimal fuss, so you can explore at your leisure and expand your horizons as you go. We'll start with acquiring the basic hardware, installing the OpenWebRX+ software, and making it fit your particular needs. If or when those needs change, you'll have the basic tools to expand your receiver to suit.

How to Use This Book

You can dip into chapters in any order you like (it's your book, after all), but each chapter builds on what has come before, so if you want to get OpenWebRX+ working quickly, then going in sequence will serve you well:

- Chapter 1, Jump Start OpenWebRX+, on page 1 guides you through buying the basic hardware needed to build the receiver, as well as installing the software.

- Chapter 2, Tame Your New WebSDR, on page 15 shows you the basics of setting up and using OpenWebRX+ so that you can listen to signals of interest to you.

- Chapter 3, Surf the Waterfall, on page 35 takes a deeper dive into the many onscreen controls available, bookmarking your favorite stations, and monitoring some of the many data signals out there.

- Chapter 4, Customize OpenWebRX+, on page 53 takes you under the hood with advanced customization, understanding regional band plans, modifying OpenWebRX+ configuration files, and more.

- Chapter 5, Explore with OpenWebRX+, on page 65 reveals the many available data decoders, reporting decoded data to online aggregators and using those services, and improving on the default reporting settings.

- Chapter 6, Share Your Decoded Data, on page 77 goes beyond listening, and explores the world of data transmissions, how to decode them, and sharing the decoded data with online aggregators.

- Chapter 7, Take Your WebSDR Public, on page 87 will show you how to be one of the hundreds of KiwiSDR and OpenWebRX+ receivers already connected to the web, so you can share what your receiver hears with online visitors.

- Chapter 8, Go Above and Beyond, on page 99 takes you beyond the limitations of the RTL-SDR receiver to help you grab more spectrum and power up your receiving experience. We'll add a few extra demodulators and explore some of the devices that live among us, quietly transmitting their data. If you're a ham radio operator, you'll also see how to use OpenWebRX+ to control your radio remotely.

Open Source Is Always Under Construction

OpenWebRX+ is an open source software project, which means you can look at the source code,[1] make copies of it, and modify it to your heart's content to add new features or to improve the way it works. You can keep those changes to yourself, or offer them back to the developers so that others can benefit. Even a very simple change can be helpful; so far, I have made one minor improvement, but I plan to do more.

This book covers OpenWebRX+ up to version 1.2.94.

Online Resources

The companion web page for this book[2] can be found on the Pragmatic Bookshelf.

I also invite you to check out my GitHub page,[3] where you'll find all the resources I have linked in the text so readers of the print edition can save typing them out by hand. There is also a lot of bonus material—things I wanted to include here that would not fit into the Pragmatic Express book format.

Up Next: Fasten Your Seat Belt

Most of us have used a traditional radio with a tuning dial and some kind of frequency display. If you've never seen or used an SDR, you may be in for a bit of shock. But it will be a good one, I promise! Let's go …

---

1. https://github.com/luarvique/openwebrx
2. https://pragprog.com/book/rmwebrx
3. https://vk2sky.github.io/

CHAPTER 1

Jump Start OpenWebRX+

Despite the phrase *software-defined* in the name, every SDR needs some hardware to do its job. So, let's start with a little bit of shopping. We'll keep it cheap and simple; we can always upgrade later if the web SDR bug really takes hold!

Every OpenWebRX+ receiver has the same basic elements:

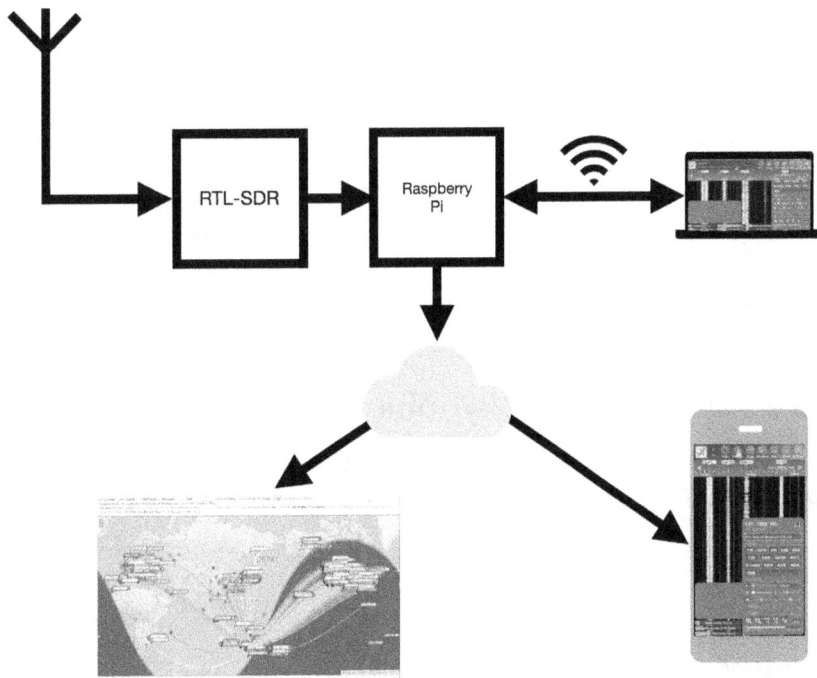

In this bird's-eye view of what we're going to build, we have the following:

- An antenna to capture radio signals

- An SDR receiver to convert the radio signals into an "IQ data stream," the *In-Phase* and *Quadrature* data that describes the received radio spectrum

- A computer running OpenWebRX+, which creates a web app and server that interprets the IQ data, presents that data as a *waterfall* in a web browser, and demodulates or decodes any signal we select. We can connect to the server over our home Wi-Fi.

- Optionally, an Internet connection to share the received audio with online listeners and decoded signals with data aggregators

Let's Go Shopping

For the SDR receiver part, we'll be using an RTL-SDR Blog V4 dongle from the RTL-SDR Store.[1] The RTL-SDR V4 can receive from 500kHz all the way up to 1.7GHz, so we'll be able to hear things like ham radio operators, AM and FM broadcast stations, shortwave broadcasters, air band voice communications, aircraft tracking beacons, and much more—even the International Space Station when it's passing overhead. This little dongle packs a lot of punch for the money!

If you'd like to get a better idea of its capabilities, YouTuber sn0ren[2] has a gentle introduction to the RTL-SDR.[3]

The RTL-SDR is a great low-cost starter SDR, but it has its limitations, mainly that it can capture only about 2.5MHz of bandwidth at a time. In comparison, for example, the more expensive HackRF[4] and SDRplay[5] models can grab 10MHz at a time, and the ready-made KiwiSDR can devour 30MHz at once—the entire LF to HF spectrum.

Capture Signals with a Skyhook

Before we can listen to any radio signals, we have to capture them! *Skyhook* is an old CB radio term for an antenna; its purpose is to "catch" radio signals.

---

1. https://www.rtl-sdr.com/buy-rtl-sdr-dvb-t-dongles/
2. https://www.youtube.com/@sn0ren
3. https://www.youtube.com/watch?v=pjoUpIIQEXk
4. https://opensourcesdrlab.com/products/clifford-r10-hackrf
5. https://www.sdrplay.com/

When it comes to antennas, the only hard-and-fast rule is "any antenna is better than no antenna." We can buy one, build one, or even improvise one from materials we have on hand. Don't sweat the details for now; keep it simple, and we can always improve things later.

If you have tried any of the online OpenWebRX receivers,[6] you may already have a good idea of the kinds of signals that interest you. Where those signals lie on the radio spectrum will determine the kind of skyhook you'll need to catch them. There is no "one size fits all" here.

Most of the signals of interest will be in either the HF or VHF/UHF portions of the spectrum. Each requires a different kind of antenna. For simplicity, we'll be starting with VHF.

VHF and UHF

Signals above 30MHz have shorter wavelengths than HF, so we can use shorter antennas. Most walkie-talkie-type radios operate in these bands simply because the antenna can be more compact.

For our initial experiments, the RTL-SDR Blog VHF/UHF dipole antenna kit is inexpensive yet versatile. It comes with two sets of telescopic antenna elements, one for VHF and one for UHF, plus tripod and suction cup mounts. You can alter the frequency coverage of each dipole by extending or collapsing the telescopic sections.

As a general guide, a dipole works best on its *center frequency*, for which you can easily calculate the antenna length.[7] The following tables show the lowest and highest center frequencies for each dipole in the RTL-SDR antenna kit.

VHF Dipole (Long Elements)

Elements	Center Frequency
Fully extended: 192cm	74MHz
Fully collapsed: 52cm	274MHz

UHF Dipole (Short Elements)

Elements	Center Frequency
Fully extended: 27cm	528MHz
Fully collapsed: 16cm	891MHz

---

6. https://www.receiverbook.de/
7. https://www.omnicalculator.com/physics/dipole

We can, of course, partially extend the elements to set the dipole length for another frequency. Don't worry too much about picking the absolute best length, though, as dipoles can pick up a broad range of signals above and below the center frequency. You can also extend the elements with extra lengths of wire to lower the center frequency if you like. These are your antennas, so feel free to experiment!

Antennas are typically horizontally or vertically *polarized* (oriented), and the polarization of the receiving antenna usually matches that of the transmitter's antenna. For example, VHF and UHF antennas on handheld radios are vertical, as are vehicle-mounted antennas for mobile stations. HF stations mainly use horizontal polarization, but verticals are often used by portable stations.

There's no hard-and-fast rule about polarization, other than "do whatever works best for you." For example, if forming the elements into a V-shaped (like the old "rabbit ears" TV antennas) or L-shaped antenna gets you better reception, then go with that.

HF

HF (high frequency) refers to the radio frequencies between roughly 3MHz and 30MHz; we also call this region *shortwave*. Signals in this part of the spectrum can travel all around the world because they can bounce (technically, they are refracted) off the ionosphere, the part of Earth's atmosphere that lies between 48km and 965km (30 to 600 miles) above sea level. This behavior strongly depends on *space weather*, driven by the activity of the Sun. Solar Ham[8] is a wonderful website devoted to the topic.

HF antennas greatly vary in size and height. For this initial example, we will use a simple *long wire* antenna, which is just what its name suggests. (The longer the better, but don't fuss too much about it for now.)

You'll find that a piece of wire does not fit reliably into the RTL-SDR's SMA antenna socket. I suggest a 1:9 HF long wire balun kit and an SMA male to SMA male cable to join the balun and the RTL-SDR. The antenna wire can then be attached to either of the contacts on the grey connector. The balun kit costs a few dollars on eBay. You'll also need an SMA to SMA cable to connect the balun to the RTL-SDR; details on that are coming up.

---

8. https://solarham.com/

Later on, we may want a more capable antenna, and there are many choices available online, such as active HF loops, multiband verticals, and so on. For VHF and up, a discone is a popular choice for its wide bandwidth. More on antennas later.

Antennas tend to work best when their size is similar to the wavelength of the signal they are catching. The lower the frequency, the longer the wavelength, and consequently, the longer the antenna needed to hear the signal effectively. HF signal wavelengths are typically between 10 and 100 meters.

An exception is radio telescope dishes, which are enormous, even though they are listening for centimeter wavelength signals. The reason is that those signals are extremely weak, so a small antenna won't capture enough raw signal for a radio receiver to make use of it. The large dish is there to focus as much incoming RF energy as possible onto the actual antenna.

Because HF antennas tend to be physically large, they often get unwanted attention from homeowner's associations and the like. If this affects you, it's worth investigating so-called "stealth antennas," a topic dear to the hearts of many ham radio operators around the world! Fortunately, wire can be easily concealed, so you probably won't experience this problem. But if you do, the *HOA Ham* channel on YouTube[9] is a great resource.

LF and VLF

For some people, these bands are the place to be. VLF is the domain of submarine communications, the Earth's natural electromagnetic rumblings, and other exotica. Antennas here can be kilometers long, so we won't deal with them in this book.

Reel in Captured Signals: the SDR receiver

Once you're catching signals with your skyhook, you'll want to reel them in and, stretching the fishing metaphor a bit, gut them as well! That's the job of the SDR.

OpenWebRX+ supports numerous SDR radio platforms[10] such as Airspy HF+ and SDRplay, which we'll come across briefly later on. These are more expensive than the RTL-SDR, but they are more powerful.

---

9.   https://www.youtube.com/@HOAHamRadio
10. https://github.com/jketterl/openwebrx/wiki/Supported-Hardware#sdr-devicesblo

Beware Fake SDRs

Before you buy an RTL-SDR, please read the Buyer's Guide at the RTL-SDR store. Many poor-quality clones out there are calling for your money, and it's easy to be fooled by an online seller. The fakes usually don't perform as well as the genuine article, and they aren't even cheaper.

More about fake SDRs and other devices can be found on the Hackaday blog.[11]

Have a look at these SDRs; one of them is not like the others:

The pretender is second from the left, a fake RTL-SDR V3. This particular one (a few years old) has a more rectangular case than the official offering, but counterfeiters are making more convincing fakes these days. Please do yourself a favor and buy from the official store.

DVB-TV (or "Digital TV Stick") Devices

According to the packaging, the older model DVB-TV dongle (far left in the previous picture) is for receiving digital TV signals. It can tune from 48.25MHz to 863.25MHz, that is, the VHF and UHF bands—but not HF. Surprisingly, these devices are still for sale online at prices similar to the much more capable RTL-SDR V4.

---

11. https://hackaday.com/2022/12/05/perhaps-its-time-to-talk-about-all-those-fakes-and-clones/

Your OpenWebRX+ Server

Next, we need a computer to run the OpenWebRX+ software.

I recommend using a small, inexpensive microcomputer that can be dedicated to being our OpenWebRX+ server. It can run *headless* (without a keyboard or monitor), so it can sit in a corner or on a shelf and not get in anyone's way.

We'll be using the popular Raspberry Pi microcomputer here. It's versatile, compact, and inexpensive, so it's a great choice for this project. (OpenWebRX+ is also available as a Docker container—see the companion repository for details.)

OpenWebRX+ can run on the low-end Pi Zero 2W, but you will get better performance with a more powerful Pi (I had a Pi 3B+ and a Pi 4 in the drawer, and have had success with both.) The more memory, the better.

If you already have a Pi, you probably also have an SD card (at least 32GB) and an SD card reader/writer; if not, you can purchase these for a few dollars.

Like any radio receiver, the RTL-SDR will work best if its antenna is outdoors (or at least close to a window) and connected with a cable that is not too long.

Optional Extras

With the antenna kit and RTL-SDR, we are good to go, but a few other modest add-ons may prove helpful.

According to the Second Law of Electronics, mechanical failures occur more often than electronic ones, so if anything is likely to break, it's usually a connector. (In case you were wondering, the First Law of Electronics is "It Works Better When You Switch It On!")

Many modern receivers, including the RTL-SDR, use SMA connectors.[12] While they are compact and perform well, they are not the most physically robust, and they can become unreliable if you disconnect and reconnect your antenna often.

It's good practice to avoid overtightening these connectors. An SMA torque wrench can help, but they can be expensive and might not be worth investing in for one RTL-SDR. If you upgrade to a more expensive SDR later, it may be worth the cost. Or, you can 3D print your own tool.[13] For now, be gentle with SMA connectors.

---

12. https://www.everythingrf.com/community/sma-connectors
13. https://makerworld.com/en/models/1045174#profileId-1030835

To reduce the wear and tear on the SMA socket, it's cheaper to attach an SMA adaptor or extension cable to your receiver and then do all the reconnecting to the other end of that. If the adaptor or cable becomes unreliable, you can replace it. It's cheaper than replacing or repairing the SDR. The RTL-SDR dipole antenna kit comes with such a cable.

Recent OpenWebRX+ updates support a GPS. Currently, this is used to automatically determine the location of your receiver and to provide an accurate time source.

Perhaps other abilities like Time Difference of Arrival (TDoA) multilateration[14] might be supported in the future. TDoA is a fancy way of saying "team up with other online receivers to work out the location of another radio transmitter." The KiwiSDR can do this.[15]

The U-blox7 (also sold under the name VK-172) is a typical compact USB GPS receiver dongle. Others, like the VK-162, have a cable and resemble a tiny wired mouse; they let you move the GPS receiver closer to the window without moving the Pi. Either kind is fine, and having a few of these on hand means you are ready for your next radio project. Be sure that you are buying a full GPS receiver, and not a GPS antenna, which would typically have an SMA plug on the end of the cable.

---

14. https://dl.cdn-anritsu.com/en-us/test-measurement/files/Application-Notes/Application-Note/11410-01009D.pdf

15. https://www.rtl-sdr.com/kiwisdr-tdoa-direction-finding-now-freely-available-for-public-use/

Install the Software

Once the hardware has arrived, we can install the OpenWebRX+ software on our Raspberry Pi.

To make things a little clearer, let's define some terms:

- *Local machine*—The computer we normally use to browse the web and download software

- *Host machine* (or simply, host)—The computer that is running the Open-WebRX+ software. We'll usually be running this machine headless, so we'll control it remotely over our network.

The easiest way to get going on the Pi is with a premade OpenWebRX+ SD card image, so let's start there.

In a web browser, go to the OpenWebRX+ Releases[16] page. The newest release of the software will be shown at the top.

Scroll down to the Assets section and locate the newest 64-bit version of the package. It will be named something like image_2025-03-01-OpenWebRX+-64bit-v1.2.78.zip, though the date and version number in the file name will be more recent. Right-click the .zip file name, click Save Link As, and save the .zip file to your local drive.

If you don't already have it, download the Raspberry Pi Imager[17] application and install it on your local machine.

Open the Raspberry Pi Imager. The startup screen looks like this:

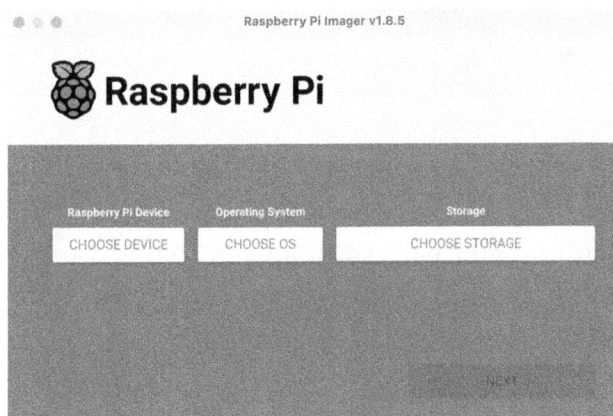

---

16. https://github.com/luarvique/openwebrx/releases/
17. https://www.raspberrypi.com/software/

We're going to write the OpenWebRX+ image we just downloaded onto the SD card, but first, we have to customize a few settings.

Initial Settings

Let's start with some basic settings to put OpenWebRX+ on our SD card:

1. Insert the card writer into a USB port on your local machine, then insert the SD card into the card writer.

2. Click CHOOSE DEVICE and select your model of Pi from the list.

3. Click CHOOSE OS, and click "Use custom" at the end of the list. Then select the OpenWebRX+ image .zip file that you just downloaded. There's no need to unzip the file.

4. Click CHOOSE STORAGE and select the SD card from the list; it will be labelled something like "Generic Mass-Storage Media - 31.9 GB."

The Imager window should now look like this:

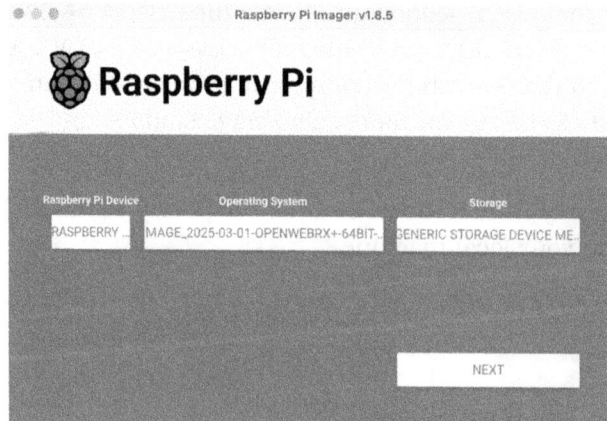

Click NEXT, and the Imager will ask if we wish to apply OS customization settings. Click EDIT SETTINGS.

The next window has three tabs: GENERAL, SERVICES, and OPTIONS.

On the GENERAL tab, we'll set our Pi's hostname, user login credentials, Wi-Fi details, and locale settings:

1. Give your Pi a unique *hostname*, the name by which it will be known on your local network. A memorable hostname is easier to remember than an IP address. You can call the host anything you like, but openwebrxplus-pi will do nicely here. Note that the Imager will add .local to the end of the hostname.

2. Next comes the username and password, which we can use to sign into the Pi, for example, "vk2sky" and "radiorocks!"; again, choose your own and make a note of them.

3. Below that, enter your Wi-Fi's name (or SSID) and password.

4. Set your "Wireless LAN country" so that your Pi uses the correct Wi-Fi channels for your geographic region. If you are unsure, consult the ISO 3166-2 list of country codes.[18]

5. Finally, select your time zone and keyboard layout from the drop-down lists. The US keyboard layout works for most of us, but some countries' keyboards have specific differences, so pick what works for you.

Next, click the SERVICES tab at the top.

Ensure that the Enable SSH checkbox is ticked, and "Use password authentication" is selected; then click SAVE. This will let you access the OpenWebRX+ server remotely later using SSH, the Secure SHell.

You don't need to change anything on the OPTIONS tab, but if you're concerned about your Pi "phoning home," you can clear the Enable Telemetry checkbox there, and again click SAVE.

Burn, Baby, Burn!

Back on the main "Use OS Customization?" screen, click YES to proceed.

We'll get one final warning that we are about to write over the SD card. Click YES again.

The Raspberry Pi Imager may ask for permission to write to the SD card; allow this, and it will then write OpenWebRX+ to the card and tell you when it's done. This process will take several minutes.

Remove the SD card from the writer and insert it into the Pi. At this point, you might as well plug your RTL-SDR into a USB port and also the optional GPS dongle if you bought one. And, of course, power on the Pi.

Add an Administrator User

We have one small task, for which we need to SSH into the Pi. If you're new to SSH, you can take a detour to Appendix 1, Crack Open the Secure SHell (SSH), on page 109 and then come back here.

---

18. https://www.ip2location.com/free/iso3166-2

Now, we're going to add an Administrator user. An Administrator has permissions to change settings via the OpenWebRX+ web interface, which we'll come to soon.

You can add several Administrators if you wish, so that others can perform administrator tasks if you are temporarily unavailable.

These are the commands we'll use shortly:

```
$ sudo openwebrx admin adduser [admin-name-goes-here]
$ sudo openwebrx admin listusers
```

For this simple example, the administrator is called "admin," but for security, you should choose a less obvious name if you are going to share your receiver online.

Tips for secure usernames and passwords are available online. You can use a tool like Bitwarden[19] to manage your usernames and passwords or follow the advice in the famous XKCD cartoon about password strength.[20]

SSH Terminal Sessions

In the SSH sessions in this book, "➜ ~" is the command prompt on my Mac's Terminal application; yours might look different.

Once we have connected to the OpenWebRX+ host machine, its prompt looks something like vk2sky@openwebrxplus-pi:~ $ but with your user and host names. The prompt may also indicate the current working directory. For simplicity, we'll show the dollar sign part of the command prompt; we don't type this in when entering a command.

Also note the .local at the end of the hostname.

You'll enter the following commands, replacing the parts in uppercase with your details:

```
➜ ~ ssh YOUR_OPENWEBRX_USER_NAME@YOUR_OPENWEBRX_SERVER
YOUR_OPENWEBRX_PASSWORD
$ sudo openwebrx admin adduser YOUR_ADMIN_NAME
YOUR_ADMIN_PASSWORD
YOUR_ADMIN_PASSWORD
$ sudo openwebrx admin listusers
$ exit
```

The session should end up looking something like this:

---

19. https://bitwarden.com/products/personal/
20. https://xkcd.com/936/

```
→ ~ ssh vk2sky@openwebrxplus-pi.local
vk2sky@openwebrxplus-pi.local's password:
Linux openwebrxplus-pi 6.6.31+rpt-rpi-v8 #1 SMP...
⋮
⋮
$ sudo openwebrx admin adduser admin
Please enter the new password for admin:
Please confirm the new password:
Creating user admin...
$ sudo openwebrx admin listusers
List of enabled users:
  admin
$ exit
logout
Connection to openwebrxplus-pi.local closed.
→ ~
```

Try This: Keep OpenWebRX+ Up-to-Date

OpenWebRX+ is open source software and still under active development, which means new features are coming out all the time. If you have the skills, you can create new features, fix bugs, make improvements to the software, and submit pull requests to make your changes available to everyone.

You can keep track of new versions and the associated changes by checking the OpenWebRX+ CHANGELOG on GitHub.[21] New versions typically get released a few days after CHANGELOG.md has been updated.

Even if you don't keep tabs on updates, you can easily download and install them at any time. Simply SSH into your OpenWebRX+ server and enter this command:

```
$ sudo apt update && sudo apt upgrade
```

Answer Y when prompted. Afterwards, enter the autoremove command to clean up any unneeded packages, and exit the SSH session:

```
$ sudo apt autoremove
$ exit
```

Up Next: Where Do You Think You Are?

With this hardware and the software installed, you should now have a functioning OpenWebRX+ receiver. But we're not done yet. The software has no idea where you are in the world, where the surrounding radio stations are, or what parts of the spectrum you are interested in. Like a newborn baby, it has a lot to learn about its world, so it's time to do some configuration. Let's get to it!

---

21. https://github.com/luarvique/openwebrx/blob/master/CHANGELOG.md

CHAPTER 2

Tame Your New WebSDR

With the basic software installation complete, we're almost ready to start playing radio! But first, we need to give the software some basic information about our receiver.

First, let's define the receiver location so that OpenWebRX+ can:

- Download bookmarks for radio stations of interest in your area, such as regional shortwave broadcasters, local amateur radio repeaters, and other OpenWebRX and KiwiSDR receivers

- Display information about regional band plans, frequency usage, and so on

- Display, on the Map page, other OpenWebRX receivers and radio stations, detected aircraft or ships, and so forth

Next, OpenWebRX+ needs to know what radio hardware we have attached and which frequency bands we want to monitor with it.

Finally, we'll update some optional settings to add a bit of personality to our receiver.

Treat OpenWebRX+ as a Web Page

Let's open our favorite web browser and open the OpenWebRX+ server, http://openwebrxplus-pi.local:8073/, replacing openwebrxplus-pi with the host name that you gave to your Pi.

The :8073 at the end of the URL is the server *port number*; if you leave it out, the browser will try to connect to the default port, :80, which is reserved for normal web servers.

Info: Why Port 8073?

You might be wondering about the port number: why :8073 in particular? As we've seen, web servers use port :80; web-related applications typically use some variation of this number, such as :8000 or :8080, sometimes with changes to the last digit. But why :8073? That probably comes from the old telegrapher's code, "73," meaning "Best wishes"; it has the same meaning among ham radio operators today.

Remember that OpenWebRX+ acts like an ordinary web page. We can change settings and click links like any other page. Nothing we do here can possibly damage the Raspberry Pi or the radio hardware, so try clicking everything! If things somehow get messed up, close the browser tab and open a new one at the same URL.

Is it Secure, and Does it Matter?

OpenWebRX (and KiwiSDR) URLs start with http:// instead of https://. This is because the software generally does not use the encrypted Hypertext Transfer Protocol Secure (HTTPS) protocol. As a result, the web browser might warn us that we are connecting to an insecure website.

Most websites these days use HTTPS, so generally, a warning about a site that's using HTTP should be heeded, especially if we're sending sensitive information such as credit card details.

In our case, though, we are connecting to a known computer on our own local network, so it's safe to click the "Continue to site" button. When we connect to someone else's web SDR, we're not sending or receiving any sensitive information (or we shouldn't be!), so there's little risk there, and we can safely connect.

However, some web SDRs have a donation button; if you click that, make sure that the donations page URL starts with https://, otherwise, some interloper could be sniffing the line for your credit card details! Credit card payments are usually handled by well-known sites such as PayPal or major credit card providers, which always use HTTPS.

Web SDRs are generally free to use, but they do cost money to run, for Internet usage, electricity, and so on. If you're a regular user of someone else's web SDR, and it has a donation button, it's a nice, friendly gesture to chip in once in a while.

OpenWebRX+ can support HTTPS, and if you have an SSL certificate, you can follow the procedure[1] described in the documentation.

Hopefully, the next thing we see is the user interface with the prompt, Start OpenWebRX+. Click it to clear the prompt and the "background fog."

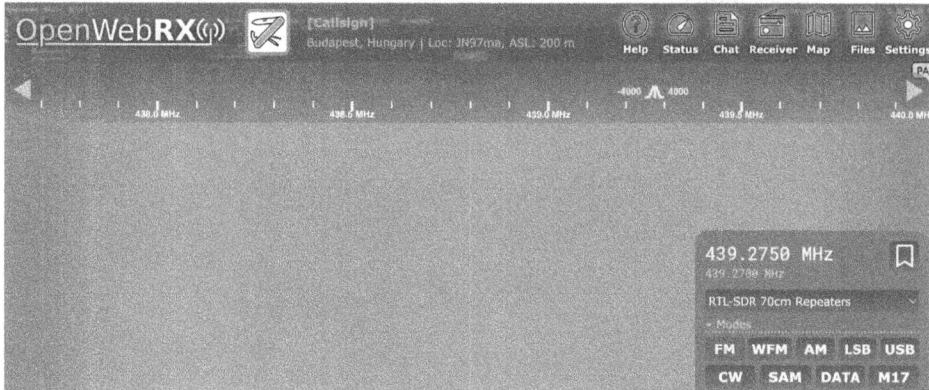

If we didn't plug our RTL-SDR properly into the USB socket (or bought one of those nasty fakes mentioned in Reel in Captured Signals: the SDR receiver, on page 5), we might see this error: "This receiver is currently unavailable due to technical issues. Error message: No SDR Devices available." Check your connections and restart OpenWebRX+ to verify that the error no longer appears.

Still Not Detecting the RTL-SDR?

If your RTL-SDR is still not being detected, SSH into the Pi, and enter the lsusb command to see all the attached USB devices. These should include the RTL-SDR, listed as RTL2838 DVB-T. If you installed the optional U-Blox GPS receiver, it will appear in the list too:

```
$ lsusb
Bus 001 Device [...] U-Blox AG [u-blox 7]
Bus 001 Device [...] Realtek Semiconductor Corp. RTL2838 DVB-T
:
etc.
:
$ exit
```

---

1. https://fms.komkon.org/OWRX/#FAQ-HTTPS

Select a Device and Band

We will explore the web interface in detail in Chapter 3, Surf the Waterfall, on page 35, but for now, we'll cover a few essentials.

Taking a closer look at the OpenWebRX+ interface, we can see the basic radio controls in the lower right-hand corner. At the top of this section is the currently tuned frequency, and below that, another frequency that changes as we move the mouse around the screen. Below that is a drop-down list labelled RTL-SDR 70cm Repeaters.

If you are familiar with one of the many KiwiSDR receivers,[2] you might have been expecting the name of a particular band, such as MW, 80m Amateur, 3MHz Aero, or similar.

In contrast, each entry in the OpenWebRX+ drop-down list shows the name of the selected SDR device, RTL-SDR, followed by the name of a *band profile*. In this case, the band profile is a section of the UHF amateur radio band, called "70cm repeaters."

As an aside, notice the Modes, Controls, Settings, and Display sections in the screenshot. There are further settings that we can hide away to reduce screen clutter; we can show or hide them by clicking the triangular icon to the left of each section label.

If you're keen to start playing, you can click the waterfall to tune in to any signals that might be about, or you can select a different band profile from the drop-down where you expect signals to be and explore there. When you click the drop-down, get ready for a little surprise.

---

2.   http://kiwisdr.com/public/

The Phantoms of the Operation

Out of the box, OpenWebRX+ comes with a number of sample band profiles to get you started. We can see several RTL-SDR entries in the drop-down list: 70cm Repeaters, 2m, and so on, all named for the wavelength (in meters) of the amateur radio signals to be found there, along with the AM Broadcast and 1090MHz ADSB for aircraft tracking.

```
✓ RTL-SDR 70cm Repeaters
  RTL-SDR 2m
  RTL-SDR 10m + CB
  RTL-SDR 40m
  RTL-SDR 80m
  RTL-SDR 160m
  RTL-SDR AM Broadcast
  RTL-SDR 1090MHz ADSB
  Airspy HF+ 20m
  Airspy HF+ 30m
  Airspy HF+ 40m
  Airspy HF+ 80m
  Airspy HF+ 49m Broadcast
  SDRPlay 70cm Repeaters
  SDRPlay 1.25m
  SDRPlay 2m
  SDRPl
```

But what's with the Airspy HF+ and SDRPlay entries? Airspy and SDRPlay are names of real SDR receivers, which we might purchase one day, but they aren't attached to our host machine right now. Let's click those entries to see what happens.

Looking at the Status panel near the bottom left of the screen, we can see that the phantom devices have failed and that OpenWebRX+ has gone in search of a working device, which is our RTL-SDR:

```
Server acknowledged WebSocket connection, OpenWebRX+ version: v1.2.93
Audio stream is compressed.
FFT stream is compressed.
SDR device "Airspy HF+" has failed, selecting new device
SDR device "SDRPlay" has failed, selecting new device
```

If we click the drop-down again, we now see that the phantom SDRs are no longer listed, and the receiver selection has reverted to RTL-SDR 70cm Repeaters. So why were those receivers shown in the first place? The answer lies in the Settings.

Disable Phantom Devices

Let's sign in as our Administrator user. Click the gear wheel icon at the top right of the screen. This opens the sign-in page in a new browser tab. Here, you can enter the Administrator name and password. When you click the Login button, you'll see the main Settings page.

There are many settings to explore here; we need only focus on a few for now.

Click SDR devices and profiles, and the Mystery of the Phantom SDRs will be resolved, as shown on the next page:

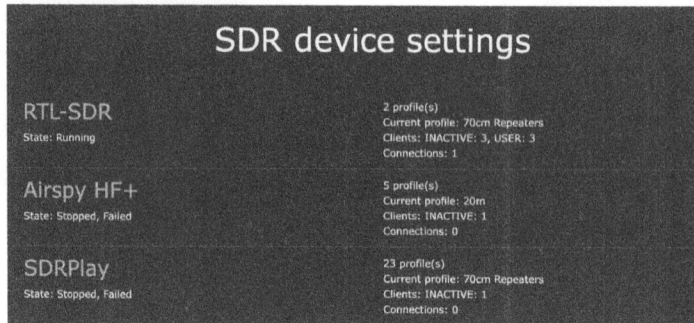

SDR device settings

RTL-SDR	2 profile(s)
State: Running	Current profile: 70cm Repeaters
	Clients: INACTIVE: 3, USER: 3
	Connections: 1

Airspy HF+ — 5 profile(s) — Current profile: 20m — Clients: INACTIVE: 1 — Connections: 0 — State: Stopped, Failed

SDRPlay — 23 profile(s) — Current profile: 70cm Repeaters — Clients: INACTIVE: 1 — Connections: 0 — State: Stopped, Failed

According to the SDR device settings, there are three SDR devices: our RTL-SDR device shows as "State: Running," which is good, but the Airspy and SDRPlay devices show "State: Stopped, Failed." This is simply because we do not have those devices plugged into our host machine.

Out of the box, OpenWebRX+ comes with these three sample device configurations to make it easier to get started with some of the more popular SDRs. The RTL-SDR configuration is useful to us right now, but the Airspy HF+ and SDRPlay, not so much.

If we restart OpenWebRX+ by refreshing the receiver browser tab, the phantom receivers will magically return to the drop-down list and will again fail if we select them, so let's deal with that problem first.

Back on the "SDR device settings" browser tab, click Airspy HF+ to view its device settings.

Simply untick the "Enable this device" checkbox, then click "Apply and save." Problem solved!

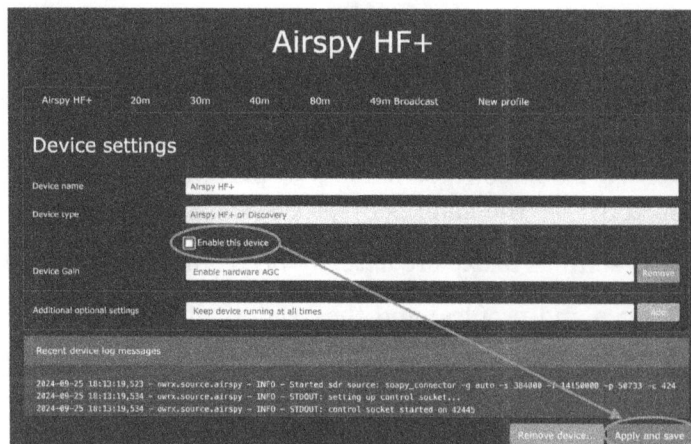

Airspy HF+

Airspy HF+ | 20m | 30m | 40m | 80m | 49m Broadcast | New profile

Device settings

Device name: Airspy HF+
Device type: Airspy HF+ or Discovery
☐ Enable this device
Device Gain: Enable hardware AGC
Additional optional settings: Keep device running at all times

Recent device log messages

2024-09-25 10:13:19,523 - owrx.source.airspy - INFO - Started sdr source: soapy_connector -g auto -s 384000 -f 14150000 -p 50733 -c 424
2024-09-25 10:13:19,534 - owrx.source.airspy - INFO - STDOUT: setting up control socket...
2024-09-25 10:13:19,534 - owrx.source.airspy - INFO - STDOUT: control socket started on 42445

Remove device... | Apply and save

If you click "SDR device settings" at the top and select SDRPlay, you can disable that phantom receiver in the same way. You can easily re-enable either device later if you wish.

There's also that big red "Remove device..." button at the bottom. We could use that instead, but let's keep those other device definitions for now; we can learn useful things from them later and enable them again if we buy one of those other devices.

Modify the RTL-SDR Device Profile

Now, let's take a closer look at our real RTL-SDR device ... again, click "SDR device settings" at the top of the page. Notice in passing that the phantom receivers are still listed, but now they are shown as "State: Stopped, Disabled."

This time, select the RTL-SDR tab:

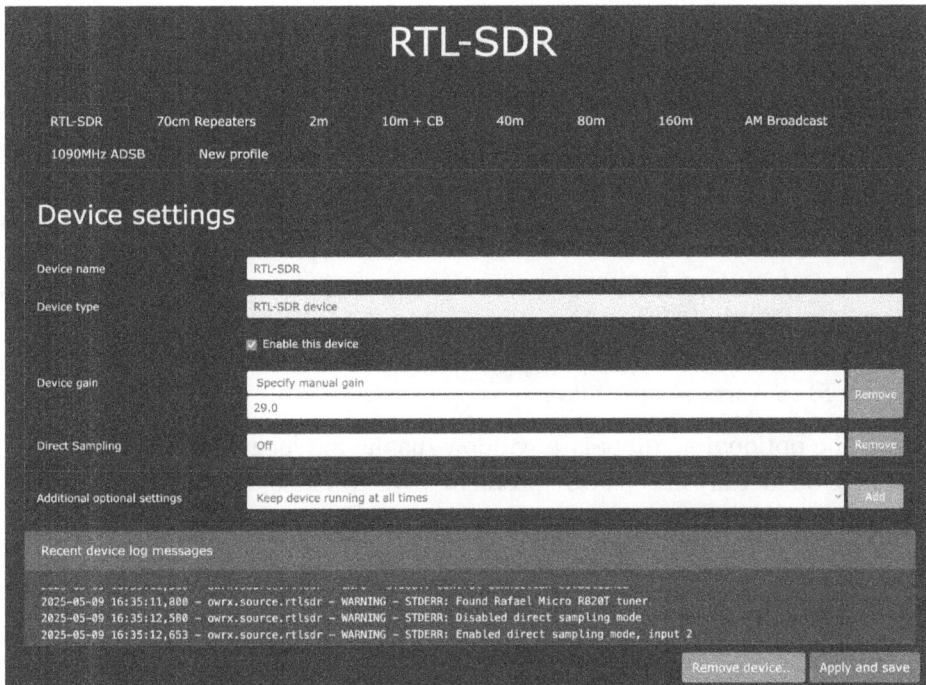

We have a number of tabs here:

- RTL-SDR—This is, of course, the device name, and this tab defines the device settings, which apply to all band profiles, though we can override them on individual band profiles if we wish.

- 70 cm repeaters, 2m, and so forth—These are the band profiles that we saw in the drop-down list earlier.

- New profile—This lets us add new band profiles; we'll use this one shortly.

The device settings for RTL-SDR are pretty simple: the device name, its type, and the "Enable this device" checkbox.

There are also device gain and direct sampling settings, which normally live in the "Additional optional settings" drop-down list below them.

You can change the device name if desired. For example, if you were to add more RTL-SDR devices later, you could rename this one to "RTL-SDR 1," and then create a second device profile called "RTL-SDR 2," and so on. More on why you would do this is coming up in Chapter 8, Go Above and Beyond, on page 99. It's best to keep the device name short to make the band profile drop-down list easier to read.

The device type is fixed, so we can't change it. But, we can always add a new device profile later, on the "SDR device settings" menu, and select a different device type then.

The "Additional optional settings" list has a number of other settings, including Device Gain and Direct Sampling, which we've already seen. These vary depending on the particular device type.

The optional settings defined on this tab are applied to all band profiles for this device. We can also apply optional settings to individual band profiles to override the corresponding settings that were set here. For now, though, we don't need to worry about them.

Below the optional settings is a window displaying logged information; this can help us identify and fix any problems. We can also copy and paste these messages to the OpenWebRX community group[3] or Discord server[4] if we need extra help.

Try This: Create a Band Profile

Let's leave the two sample profiles aside for the moment and add one of our own. FM broadcast radio is popular worldwide, and most of us have a station somewhere nearby, so let's create a new profile to capture a local FM station. Alternatively, we can use an AM broadcast station instead.

---

3. https://groups.io/g/openwebrx
4. https://discord.com/invite/gnE9hPz

In either case, make sure that you have attached one of the antennas we have previously discussed, for example, the VHF/UHF dipole kit for FM, the long wire for AM, or some other suitable antenna.

In most parts of the world, 76–108MHz or 88–108MHz is dedicated to FM broadcasting. That's a 20MHz (or more) chunk of radio spectrum, which is far too much for an RTL-SDR to swallow all at once. Let's start with a section that an RTL-SDR can digest.

According to the RTL-SDR blog,[5] the device can take up to 3.2MS/s (million samples per second), which is equivalent to 3.2MHz of radio bandwidth, but it's unstable at that rate. The blog recommends a maximum of 2.56MS/s, so let's be a little bit more conservative and sample at 2.5MS/s. We can experiment with higher rates later if we like.

Another factor to consider is the amount of processing power your host machine has: a Raspberry Pi 4 or 5 will easily cope with processing that much receiver data, but a lower-end Pi Zero W might struggle to keep up. If it does, you might have to settle for a lower sampling rate and thus a smaller chunk of radio spectrum. You'll learn how to check how well the host is coping in Chapter 3, Surf the Waterfall, on page 35.

So, let's pick our favorite local station, grab 2.5MHz of spectrum that includes its frequency, and see how we go. For this example, we'll pick ABC Classic FM, a Sydney station that broadcasts classical music on 92.9MHz.

We will center the 2.5MHz slice of spectrum on 93.0MHz so that the preferred station will be close to the middle. This gives a nominal coverage of 92–94MHz (a nice round 2MHz wide), plus an extra 250kHz overlap at either end; this is useful for covering any stations close to the edges of the 2MHz slice. If trying this on the AM broadcast band (also known as the Medium Wave or Medium Frequency band), pick a center frequency of 1.5MHz, which will cover 250kHz to 2.75MHz, more than enough for the entire AM band.

Next, click the "New profile" tab, and fill in the form like this:

- Profile name—This profile is called "FM Broadcast 92-94," but you can call yours whatever you like (but again, keep it short, as this will appear in the drop-down list after the device name).

- Center frequency—First, use the drop-down at the right to select "MHz," then enter a frequency right in the middle of the desired band (tip: if you enter the number first and then select the units, the number will get

5. https://www.rtl-sdr.com/about-rtl-sdr/

scaled accordingly, which can be confusing). This example uses 93MHz, but a decimal value like 92.9 is perfectly fine.

- Sample rate—First, pick "MS/s" from the drop-down at the right, then enter 2.5 in the text box.

- Initial frequency—This can be anywhere in the part of the spectrum you are capturing. Let's go straight to ABC Classic FM. So, again select "MHz" from the drop-down list, and enter 92.9 or the frequency of your local FM station.

- Initial modulation—This selects the usual type of signal modulation in this band. For FM broadcasting, that's *wideband FM*, so select WFM from the drop-down. For AM broadcasts, the initial modulation is, unsurprisingly, AM.

- Tuning step—This selects how much the frequency should change when you tune up or down the band. FM broadcast stations are typically multiples of 100kHz apart. The highest available tuning step is 50000 Hz (50kHz), so pick that one from the list. For AM, the tuning step is usually either 10kHz (in the Americas) or 9kHz (in the rest of the world).

- Additional optional settings—Keep it simple and disregard these for now. You can always modify the profile later.

Finally, click the blue "Apply and save" button at the bottom right. Your new profile should look something like this:

RTL-SDR

| RTL-SDR | 70cm Repeaters | 2m | 10m + CB | 40m | 80m | 160m | AM Broadcast |

| 1090MHz ADSB | FM Broadcast 92-94 | New profile |

Profile settings

Profile name	FM Broadcast 92-94
Center frequency	93	MHz
Sample rate	2.5	MS/s
Initial frequency	92.9	MHz
Initial modulation	WFM
Tuning step	50000 Hz
Additional optional settings	Device gain	Add

Settings / SDR device settings / RTL-SDR / FM Broadcast 92-94

Move up Move down Clone Remove profile... Apply and save

Adding band profiles is as simple as that. You can add as many profiles as you like, for whatever bands your device can receive. You can also use the Clone button to copy the details from the selected band profile to the New Profile tab.

Let's See What We Can Hear

Now, let's check out how well our new band profile works. If we go back to the OpenWebRx+ receiver tab in our browser and click the drop-down list, we can see our new receiver/band profile at the end of the list:

Select the new band from the drop-down, and let's have a look at the waterfall:

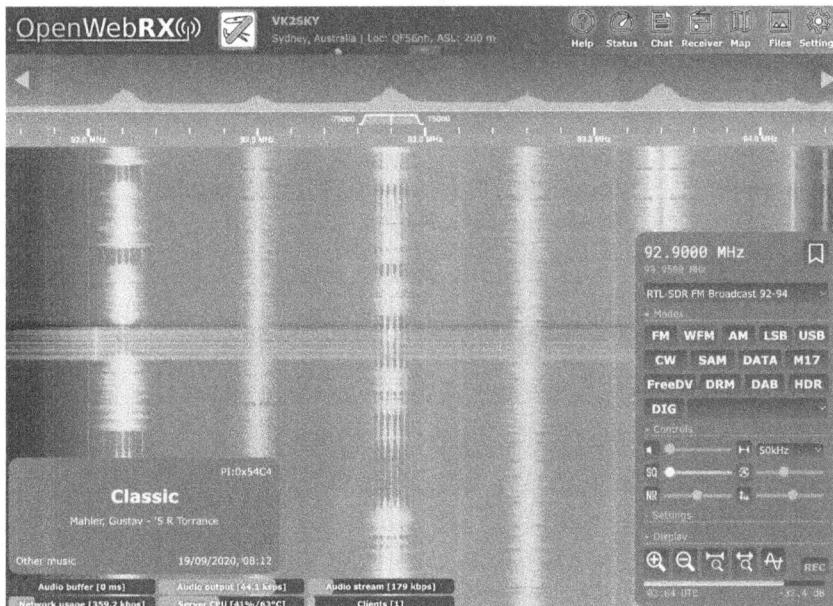

The waterfall reveals stations in the mostly yellow and red vertical bars against the blueish background. Sure enough, ABC Classic FM[6] is there in the middle. There's also a nasty-looking horizontal stripe, halfway down the waterfall; this usually indicates interference, which, in this case, lasted only a few seconds.

In Sydney, several other stations live in this part of the FM broadcast band, among them 2MFM Community Radio[7] on 92.1MHz, and Koori Radio[8] on 93.7MHz.

Clicking a vertical stream tunes into the associated station.

Spectrum Scope

Just above the frequency bar is the spectrum scope, a view of station activity right now. It's hidden by default, but you can switch it on and off by clicking the "sine wave" button near the bottom right of the receiver controls panel, beside the REC button.

Selecting ABC Classic FM uncovers another little detail: the box near the bottom left, just above the Status panel, displays Radio Data System[9] information, such as station identification, details of the music being played, and so on. Here, ABC Classic FM was playing Gustav Mahler's Symphony No. 5, and Russell Torrance was presenting the program.

This briefly demonstrates one of the other great features of OpenWebRX+. As well as listening to stations, we can also decode data that they might be transmitting. We'll investigate this further in Chapter 5, Explore with Open-WebRX+, on page 65.

Radio Data System in the USA

If you are in the USA, and the RDS data box is empty, go to Settings > Demodulation and decoding > Miscellaneous, and tick the "Decode USA-specific RBDS information from WFM broadcasts" checkbox.

If the RDS data box is still empty when an FM station is tuned in, the station is either not broadcasting RDS data or the signal is too weak to decode properly. Using a better antenna may help.

---

6. https://www.abc.net.au/listen/classic
7. https://www.2mfm.org/
8. https://kooriradio.com/
9. https://www.sigidwiki.com/wiki/Radio_Data_System_(RDS)

Try This: Add More Band Profiles

Now that you've successfully added a band profile, it will be easy to add more.

If you want to cover the entire FM broadcast band, you can add more band profiles like the one you just created. These are the only things you would need to change on the new profiles:

- Profile name—A unique name displayed on the drop-down list

- Center frequency—For example, to cover 94-96MHz, you would select a center frequency of 95MHz

- Initial frequency—The frequency of a station of interest

The other settings would be the same as your first FM broadcast band profile.

Assuming you divided the FM band into 2MHz-wide segments (using the 2.5MS/s sample rate to provide a little overlap to accommodate stations on the edges), you would require 10 such profiles for the RTL-SDR to cover 88–108MHz. Or, you could start thinking about more powerful SDR receivers that can capture more radio signals at once. For now, 2MHz segments are fine.

Be Careful What You Listen For

A word of advice: just because we can make OpenWebRX+ listen to any frequency the SDR can hear, doesn't mean that we should. In some countries, it's illegal to listen to certain transmissions, such as police, military, and even air traffic control. In other places, you may listen, but you may not tell others about what you've heard.

Check the applicable laws where you are or where you intend to travel to before tuning in. Depending on where you are, possessing any kind of radio scanner can get you into trouble. As much fun as it is to listen to all kinds of radio signals, it's not worth going to jail for it.

Try This: NOAA Your Weather

In the United States, a great service to monitor is NOAA Weather Radio,[10] which provides 24-hour-a-day automated transmissions from the National Weather Service.

---

10. https://www.weather.gov/nwr/

Create a new band profile with these settings:

- Profile name—For example, "NOAA Weather Radio"

- Center frequency—162.475MHz

- Sample rate—250kS/s (note "kS/s" instead of "MS/s," because the NOAA stations occupy a narrow slice of the spectrum)

- Initial frequency—Set this to the frequency of your nearest NOAA station, or the center frequency

- Initial modulation—FM (narrow-band FM, not WFM used by music broadcasters)

- Tuning step—25000 Hz

Finally, click "Apply and save."

Once OpenWebRX+ has had a chance to download band data files, the channels will be automatically labelled NOAA-1 through NOAA-7.

Try This: Sort Your Band Profiles (That's an Order!)

We can also change the order of the band profiles on our drop-down list, perhaps to put our favorite profiles near the top, or to make sure that the profiles are in frequency order, and so on.

Go back to the RTL-SDR settings, select the profile of interest, and click the Move Up or Move Down buttons at the bottom to change the order. No need to click "Apply and save." When you go back to the main receiver tab, the profile order in the drop-down will have changed.

Fine-Tune the Receiver Settings

The next most useful thing we can do right now is to update some of the general settings for our receiver. Let's click Settings at the top, then the General Settings button.

We'll go through the essential settings next.

Put the Receiver on the Map

First, let's change the map. Scroll down to "Map settings," and change the map type from Google Maps to "OpenStreetMap, etc." We'll need this shortly when we talk about Maidenhead Locator squares. Okay, scroll back up to the top.

The "Receiver information" section mainly deals with the receiver's geographical location; it also determines how the receiver's entry will appear in the Receiverbook directory[11] if we later decide to list ours online.

- Receiver name—First, you can give your receiver a name. If you have an amateur radio call sign, it helps to include that in the name, but that's up to you. Calling it "My Spiffy OpenWebRX+ Receiver" is perfectly acceptable. This name gets displayed at the top of the receiver page and also appears in the online receiver directories.

- Receiver location—This is the name of your location, which gets displayed beneath the receiver name.

- Receiver elevation—This is roughly how far above sea level the receiver is, in meters. This can help visitors judge the radio quality of the receiver location. For example, a VHF receiver on a mountain will usually hear signals better than one in a valley.

---

11. https://www.receiverbook.de/

- Receiver admin—This is a contact email for questions about your receiver, if you wish to publish one.

- Receiver band plan—The upper and lower limits of bands depend on which of the world's three International Telecommunications Union regions[12] you are in, so pick the one that includes your location. Band plans can vary within each ITU region, so we may need to fine-tune the band plan later. We'll see how to do that in Lord of the Files, on page 56.

- Receiver coordinates—This is the reference location for your receiver. The coordinates do not get displayed to users of the receiver, though the six-character Maidenhead Locator gets displayed at the top.

If you're concerned about security, simply drag the map so that the red marker is over a nominal location for your receiver. OpenWebRX+ uses this location to identify and display nearby transmitters and receivers on your map, so it should be reasonably close to your actual location.

> At the time of writing, even though we changed the map type to OpenStreetMap, the setting doesn't get applied to this particular map.

If you have a GPS dongle attached, you can use that instead of manually entering your location: tick the "Update receiver coordinates via GPS (if device present)" checkbox.

OpenWebRX+ also uses our location to determine what transmitters, repeaters, and other webSDRs are in our area, so it can display them on the map. It also automatically sets the band definitions for our region.

Try This: Explore the Maidenhead Locator Grid

Click the Map icon at the top of the OpenWebRX+ screen, which will open a new browser tab, and tick the Maidenhead-QTH checkbox in the map legend at the bottom left. Zoom in and out, and notice how the grid square identifiers change. Find your own Maidenhead locator, to different degrees of precision.

We can also enter a street address into WhatsMyLocator[13] to find the corresponding grid square.

Finishing off the Receiver Information section:

---

12. https://www.itu.int/en/ITU-R/information/Pages/emergency-bands.aspx
13. https://www.whatsmylocator.co.uk/

About Maidenhead Locators

Amateur radio operators often like to exchange information about where they are, for example, to keep track of the places they've contacted, to know which way to point a steerable antenna to get the best signals to and from the other station, and so on.

But conventional latitude and longitude are cumbersome to use, especially if sending them over a noisy radio link. So, a shorthand method like the *Maidenhead Locator System* comes in handy.

The system divides the world into a number of "grid squares," which are specified by a pair of letters. Each square is divided into subsquares specified by a pair of digits, and so on, alternating between letters and digits. The first of each pair refers to the longitude, the second to the latitude. The more pairs you use, the more precise the square. Most radio hams use four or six-character locators.

Examples:

QF = the state of New South Wales in Australia, an area a little bigger than Texas (sorry, Texan friends, but it's true!)
QF56 = the Greater Sydney and Blue Mountains region
QF56OD = Sydney's central business district
QF56OD54 = the south end of Sydney Harbor Bridge to Bennelong Point
QF56OD54UI = the Joan Sutherland Theater at the Sydney Opera House

- Photo Title and Photo Description—Many OpenWebRX+ owners like to personalize their receivers with a photo, such as a local landmark, national symbol, or similar iconic image. The photo and these two fields are then displayed when the user clicks the middle of the top section of the receiver window, to the left of the Help icon. You can specify the images in the next section.

Put Your Receiver in the Picture

These settings add a bit more personality to our receiver, but we can ignore them:

- Receiver Avatar—This is the icon that appears in our receiver's Receiverbook entry; see the Receiverbook directory for other examples.

- Receiver Panorama—This is the photograph at the top of the receiver window, along with the title and description mentioned in the previous section.

Receiver Limits and Receiver Listings

We'll come back to these in Chapter 6, Share Your Decoded Data, on page 77.

Paint the Waterfall

Scrolling down, we can customize the waterfall. We can leave these at their default values, but two of the settings are worth noting:

- Waterfall color theme—We can change this setting to suit our own tastes. Try out the various theme options and see how they look on the waterfall. Try the Eclipse theme, for its more vivid colors. There is also a custom setting, so we can roll our own, for example, to make it easier for colorblind users.

- Automatically adjust waterfall level by default—It's a good idea to tick this checkbox, so that the waterfall colors stay reasonably consistent under varying band conditions. For example, the shortwave bands can be full of static one day, then be quiet the next, making the colors of the "no signal" parts of the waterfall change if the software doesn't automatically adjust. We can enable this behavior for all devices by ticking the checkbox here, or do it for particular SDR devices in their settings, or for individual band profiles on each profile's optional settings.

Compression and Display Settings

The Compression settings can be left at ADPCM, which is Adaptive Differential Pulse Code Modulation.

Let's look at the Display settings:

- Tuning precision—We can set a default that will apply to everything, and override it by device or band. 100Hz is a good starting value.

- Shortwave bookmarks range—Here's one of the places where the receiver location comes into play. OpenWebRX+ queries the EIBI radio stations database[14] so that it can add labels above the waterfall for stations within the specified range. This makes it easier to identify stations on the receiver without manually searching radio station frequency listings.

---

14. http://eibispace.de/

Home in on the Range
If we set the shortwave bookmark range too high, our display may get cluttered with markers, including for stations that we can't hear; too low, and we may miss labelling stations that we can hear. Shortwave signals can travel thousands of kilometers, so try different values and see what works best for you.

- Repeater bookmarks range—This also uses the receiver location as a reference point. For ham radio operators, OpenWebRX+ searches the RepeaterBook[15] database for amateur VHF and UHF band repeaters within the specified range. 100km is a reasonable starting value. Clickable labels for all repeaters in range are displayed in yellow above the waterfall.

Plotting Received Data

OpenWebRX+ can display decoded data on a map; for example, ADS-B[16] aircraft position beacons or AIS[17] shipping transmissions.

Displaying this information usually involves consulting online databases to retrieve aircraft or ship information. We'll come to these in Chapter 5, Explore with OpenWebRX+, on page 65, but for now, the default URL settings here will suffice.

If you have changed anything here, remember to click "Apply and save"!

Up Next: Let's See What This Baby Can Do

That was a lot of configuring! We covered the main general settings for the receiver, and we created a few band profiles that let us listen to parts of the spectrum that interest us. We can add, change, and remove these profiles any time we like.

We also briefly looked at the Maidenhead Locator System, a location shorthand used mainly by ham radio operators, and had a brief look at the Map page, which we can use for plotting the locations of aircraft, ships, APRS stations, and more.

That's enough to get us going. There are more settings we could change now, but where's the fun in that? Let's spend a bit of time actually *using* the receiver.

---

15. https://www.repeaterbook.com/
16. https://www.sigidwiki.com/wiki/ADS-B
17. https://www.sigidwiki.com/wiki/Automatic_Identification_System_(AIS)

CHAPTER 3

Surf the Waterfall

Now that we've done some basic configuration, we can start receiving some real radio signals. Right then, let's play with this thing!

With a traditional communications receiver, we switch it on, turn the tuning dial to select a frequency, and then pick an appropriate demodulator or mode.

With OpenWebRX+, we pick a frequency, either by clicking the waterfall or typing in the desired frequency; then, we select an appropriate demodulator.

From our ABC Classic FM example earlier, we were able to pick the band profile that included the station, enter the station frequency (92.9MHz), select the WFM demodulator, and listen to music, sweet music!

But radio is a lot more than listening to local broadcasters. Unusual stations are waiting to be discovered, and there are exotic transmission modes to explore, "secret" data to uncover, and more, so let's go a little deeper.

You might like to tune in to one of your favorite stations to keep you focused while you work through the details.

Go over the Waterfall (No Barrel Required)

The most noticeable feature of OpenWebRX-based receivers is the *waterfall*, a representation of the captured radio spectrum (along the horizontal axis) over time (on the vertical axis). We can see many stations cascading down the waterfall, and click any of them to tune in.

The color of each pixel indicates the strength of the received signals at a given frequency and time. In the default color scheme, signal strength follows the rainbow: red signals are the strongest, yellow and green are mid-strength, the weakest signals and background noise are shades of blue.

The top of the waterfall shows what has happened most recently, while the bottom shows how things were about 30 seconds ago; the exact amount of time depends on the waterfall configuration settings.

The waterfall is especially useful to observe radio signals that come and go, such as two-way radio conversations and bursts of data; in contrast, broadcast stations have a continuous signal, so they are easy to spot at any time.

Enter the (Frequency) Matrix

In the 1999 cyberpunk action film, *The Matrix*, Cypher introduces Neo to the Matrix code, a cascade of symbols (mostly stylized katakana)[1] representing on screen the activity going on inside the digital world, commenting "you get used to it: I don't even see the code—all I see is blonde, brunette, redhead ...."

The waterfall is a bit like that. The colored vertical strips show signal activity on each frequency, and with a bit of practice, we can tell what's going on by looking at these strips. After a while, we can tell what kinds of signals are there, without having to listen to them:

- AM and FM broadcast stations appear as continuous strips, as they transmit non-stop. The carrier frequency is usually the strongest part of the signal, appearing in red down the middle, with yellow or greenish edges showing the upper and lower sidebands with the audio.

- Two-way radio signal strips stop and start as the operators take turns talking. If they are using a single sideband mode, the strongest part of the signal is usually on the left (for the lower sideband) or right (for the upper sideband).

- Morse code signals also start and stop; we can even see the "dots" and "dashes" if we rotate our head anticlockwise.

And so on. The online Signal Identification Guide[2] is the place to go to see the huge variety of these "waterfall signatures," to hear samples, and to learn about how and where these modes are used.

As we saw when we set up and tested our first band profile, clicking these vertical stripes tunes to the station occupying that frequency.

---

1.   https://www3.nhk.or.jp/nhkworld/lesson/en/letters/katakana.html
2.   https://www.sigidwiki.com/

The Art of Noise Reduction

In Paint the Waterfall, on page 32, we configured the waterfall to automatically adjust its colors to adapt to the background noise level. Using the FM broadcast band profile, we can see this in action when we refresh the receiver page:

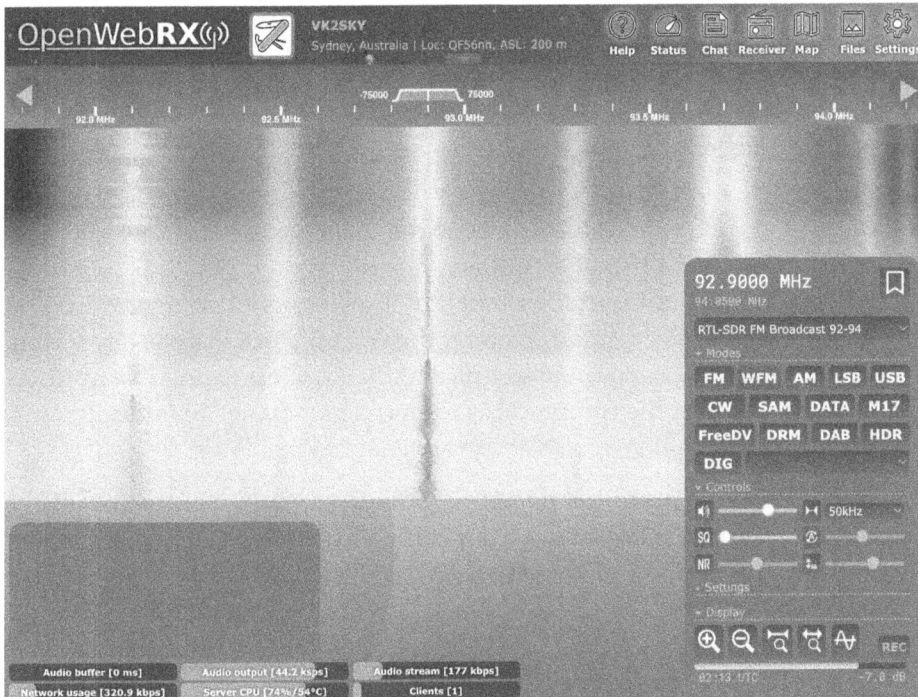

Notice how the background color at the bottom of the waterfall (the "oldest" part) is an intense red (very strong signal) on a greenish yellow (mid-strength signal) background. Closer to the top of the waterfall, the red has reduced to reddish yellow, and the background is blueish, indicating little or no signal.

In fact, there are only three signals real here, indicated by the green trace at the top left on 92.1Mhz and the stronger (red) signals on 92.9 and 93.7MHz.

The other vertical traces could be "image frequencies," usually caused by strong signals overloading the SDR receiver. Images can sometimes be lessened by reducing the sample rate or making the antenna smaller so it is less sensitive. If you live close to the transmitters, sometimes you have to live with the effect.

Master the OpenWebRX+ User Interface

Let's take a look at the other parts of the OpenWebRX+ user interface. OpenWebRX+ has a large number of controls and settings, which can seem overwhelming at first, but we don't need to know them all straight away. Most of the time, we can simply pick a band profile, choose a frequency by clicking a station, and choose the appropriate demodulator.

Let's start at the top:

- Clicking the OpenWebRX logo opens the main OpenWebRX website in a new browser tab. The site contains documentation (for the "non-plus" version, OpenWebRX, though much of it also applies to OpenWebRX+), links to the Receiverbook[3] directory, the OpenWebRX community forum, and more. If you encounter technical difficulties operating OpenWebRX or OpenWebRX+, the community forum is a great place to find solutions. There's also an OpenWebRX Telegram channel.[4]

- Avatar image—This is the image we defined in Put Your Receiver in the Picture, on page 31; clicking it (or anywhere between the avatar image and the Help button to the right), toggles the Panorama image that we defined along with the avatar.

- Continuing to the right, we have the Receiver information we configured earlier:

    - Receiver name
    - Location name
    - Maidenhead grid square (to six-character resolution)
    - Height above sea level
- The Help button opens the OpenWebRX+ documentation in a new browser tab; this documentation covers the many extensions and improvements specific to the "Plus" version of the OpenWebRX software.

- The next three buttons toggle parts of the screen on and off; this can be useful to reduce screen clutter. Try each of them in turn and note the effects. We'll come back to these individual panels shortly:

3.　https://www.receiverbook.de/
4.　https://t.me/openwebrx

    - Status
    - Chat (or Log, if chat between users is disabled)
    - Receiver
- Map—Opens the map view in a new browser tab. We'll return to the Map in more detail when we come to plotting received data in Chapter 5, Explore with OpenWebRX+, on page 65.

- Files—Some data modes, such as Slow Scan Television (SSTV), radio facsimile (FAX), and others, create local files so that the images or data can be reviewed later. Clicking the Files button opens the review page in a new browser tab.

- Settings—As we saw earlier, when we first configured OpenWebRX+, Settings opens the Administrator login page in a new tab.

Next, we have the Frequency Scale, which shows the range of frequencies being displayed in the waterfall:

We can also see several yellow labels, which indicate particular "spot" frequencies of interest. In the previous image, the receiver is tuned to 147.000MHz, which in the Sydney area is VK2RWI, a popular amateur radio repeater. We can also see VK2ROT, another repeater closer to the city center, and several others, whose call signs are a little too long to fit on their labels.

These markers show frequency bookmarks automatically downloaded each day from either RepeaterBook,[5] or the EiBi Shortwave Broadcaster Database.[6]

Let's have a closer look at VK2RWI:

Below the VK2RWI label, and above its frequency, is a graphical indicator showing the passband of the tuned signal. The passband is the narrow slice of radio spectrum that we are focused on when tuned to a particular frequency.

---

5. https://www.repeaterbook.com/
6. http://eibispace.de/

The small vertical bar in the middle marks the carrier or dial frequency of the signal; in other words, it shows the nominal frequency we have selected. To the left and right of this bar, the lower and upper sideband edges are marked with the numbers -4000 and 4000, indicating a total passband width of 8000Hz or 8kHz. So, when we have tuned to "a frequency," we have in fact selected a small slice of the spectrum.

Whether a received signal shows anything in the carrier or sideband areas depends on the type of signal. VK2RWI is a narrow band FM repeater, and FM signals here have all three elements. The sidebands are typically 3kHz (or 3000 Hz) wide, so the 4kHz upper and lower passbands are wide enough for the 6kHz-wide signal here to be received clearly. A radio signal can contain all of these, or just one. For example, a USB signal contains only the upper sideband—no carrier or lower sideband.

Narrow band FM, often referred to as NBFM, is typically used for two-way voice communications, such as walkie-talkies or mobile radio. In contrast, wideband FM, used for high-quality music broadcasting, is typically 100kHz wide. OpenWebRX+ uses a 150kHz passband for these signals.

Modifying the Passband

We can use the mouse to drag the edges of the passband, for example, to reduce the passband "skirts" and minimize interference from signals nearby on the waterfall. To restore the passbands to their default settings, press the "|" key.

Occasionally, we'll see green labels; these indicate the operating mode typically used on the marked frequency, according to the regional band plan file or user conventions. For example, 7.074MHz in the amateur 40m band is typically used for FT8[7] mode communications.

We can also set "personal" labels, which get displayed in blue; these are saved in the browser's local storage area, and are not visible to other users or even to other web browsers on the same computer. We'll set one of these shortly.

At each end of the frequency scale, you'll see a triangular icon. Click these to tune up or down by the current tuning step. Note that if we tune up or down beyond the displayed edges, the waterfall does not scroll horizontally to keep the tuning indicator on screen. We can use the "Zoom in totally" receiver control button to recenter the waterfall on the tuned frequency, or we can simply drag the waterfall until the tuning indicator is visible again.

---

7.   https://www.sigidwiki.com/wiki/FT8

The Status panel provides various statistics related to our OpenWebRX+ server; we can turn it on and off using the Status button at the top right of the screen.

In the following example, the OpenWebRX+ server (a Raspberry Pi 3B+) is monitoring the repeater section of the 2m VHF amateur band:

Audio buffer [0 ms]	Audio output [44.2 ksps]	Audio stream [45 kbps]
Network usage [194.1 kbps]	Server CPU [25%/63°C]	Clients [2]

Note that the Pi is cruising along at about 25 percent capacity and 63 degrees C (145 degrees F).

For comparison, here is the same Raspberry Pi, capturing a different section of the 2m VHF amateur band but also decoding three packet/APRS channels in the background and reporting decoded data to the local APRS-IS[8] server. Notice that the Server CPU load and temperature are both higher than in the previous example:

Audio buffer [0 ms]	Audio output [0.0 ksps]	Audio stream [0 kbps]
Network usage [144.4 kbps]	Server CPU [88%/71°C]	Clients [2]

Incidentally, the Audio output and Audio stream values are at zero: OpenWebRX+ turns off the sound because packet radio signals sound rather harsh.

For voice modes, we can expect these numbers to typically be in the tens of kilosamples or kilobits per second.

If the Audio buffer stays high, it may indicate that the Pi is having trouble keeping up with the received audio. We can reduce the load, for example, by specifying a lower sampling rate, turning off some of the background decoding, or by installing a higher-spec Pi.

"Network usage" indicates the data rate from the OpenWebRX+ server on our local network. Even if we are the only person connected to the server, we can expect a rate of several hundred kilobits per second. If you're sharing your receiver with external users, we could expect this figure to rise, so keep in mind your data plan to avoid billing surprises.

"Clients" shows the number of users connecting to the OpenWebRX++ server. If the Server CPU load goes red when the number of clients is high, you might want to consider lowering the maximum number of clients in Settings >

---

8.   https://www.aprs-is.net/

General settings > Receiver limits from its default value of 20, or you could upgrade to a more powerful Raspberry Pi.

All of these figures tend to fluctuate from moment to moment, so don't lose too much sleep over them. Of course, we can switch off the Status panel to reduce screen clutter.

The Log panel serves a dual purpose: 1) displaying system messages, and 2) letting users send short messages to each other, for example, to share news of interesting signals. The user chat option can be disabled in Settings > General settings > Receiver limits. Clear the "Allow users to chat with each other" checkbox. This panel can also be hidden, using the Log button at the top right of the screen.

Try This: Chat with Other Users

At the moment, we're the only one using the OpenWebRX+ server, so there's nobody to chat with, but let's pretend for a moment. Notice the text entry fields labelled Name and Message in the Chat panel; we'll use those in a moment. Also, notice "Clients [1]" in the Status panel at the bottom of the screen:

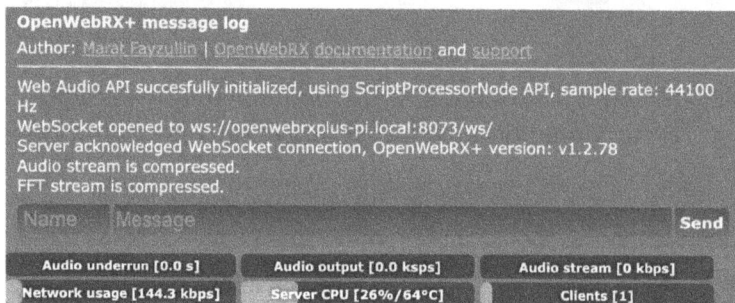

Now let's open another browser tab and open another OpenWebRX+ receiver there. The Status panel now shows "Clients [2]." Back in the Chat panel, we can enter a user name and a message, and click the Send button:

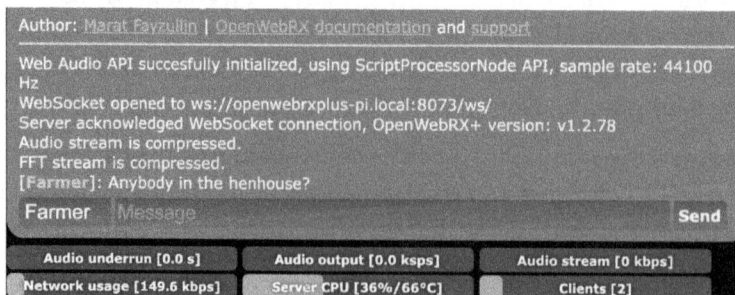

Back in the first tab, that message is visible, and we can respond in the name of the other user:

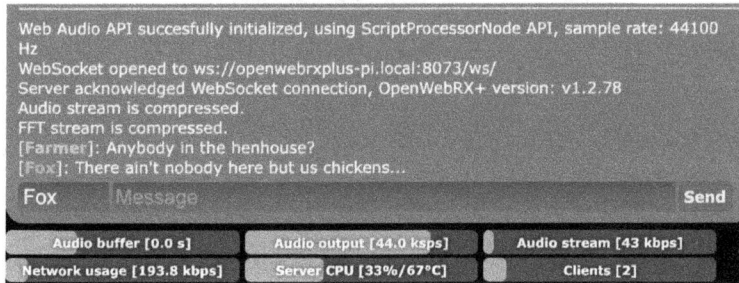

Later, if we open the receiver to other online listeners, we can have real chats about what we're hearing on the radio, or about other topics such as Blues musician Louis Jordan,[9] in case you were wondering about those chickens.

Try This: Update OpenWebRX+

When OpenWebRX+ starts up, it displays the current version number in the Chat window:

Reminder: it's worth updating the software regularly to take advantage of new features, bug fixes, and so on. The procedure is simple: SSH into your OpenWebRX+ server and enter the command:

```
$ sudo apt update && sudo apt upgrade
```

Follow the prompts, and when any updates are complete, use the exit command to close the session.

OpenWebRX+ should detect any relevant changes and restart the server if necessary. We'll usually see a few "Websocket has closed unexpectedly. Attempting to reconnect ..." error messages while the server restarts, but it should soon settle down and display the updated version number:

---

9. https://www.youtube.com/watch?v=r8RBZNWXxZE

```
WebSocket has closed unexpectedly. Attempting to reconnect in 4 seconds...
WebSocket error.
WebSocket has closed unexpectedly. Attempting to reconnect in 8 seconds...
WebSocket error.
WebSocket has closed unexpectedly. Attempting to reconnect in 16 seconds...
WebSocket opened to ws://openwebrxplus-pi.local:8073/ws/
Server acknowledged WebSocket connection, OpenWebRX+ version: v1.2.79
Audio stream is compressed.
FFT stream is compressed.
```

Next up is the Receiver Control panel at the bottom right, which we can show
or hide with the Receiver button. The panel itself has several sections, which
we can hide to simplify the display. Click the small triangle icons beside the
Modes, Controls, Settings, and Display headings to collapse those sections
so that the Receiver Control panel looks like this:

Try This: Different Tunes

We've already seen how we can simply click the waterfall to select a station,
but it's not the only way to tune the receiver.

At the top of the Receiver Control panel, we can see two frequency numbers,
a "bookmark" icon, and the device/band profile drop-down that we met earlier.

The upper frequency number (92.9000MHz) is the currently tuned frequency;
this changes every time we click the waterfall.

Tune Directly with the Keyboard

We can also manually tune to a specific frequency. Click the frequency
number and it becomes a text entry field with a units selector: Hz, kHz, MHz,
GHz, and THz. We can use the text field to jump directly to the frequency of
interest.

As we saw when setting up profiles, it's best to select the units first, then enter the number. Here, we're in the 92–94MHz segment of the FM broadcast band, so let's select MHz from the drop-down list.

Type the desired frequency, and press Enter to confirm it. If we enter a number outside the range specified in the band profile, it will be ignored.

We can also tune up and down by hovering the mouse over a digit of the displayed frequency and rolling the mouse wheel. How fast we tune depends on which digit we hover over. Or we can use the cursor Up and Down keys to modify the frequency by 1MHz.

Tune with the Mouse

Next, if you move your mouse around the waterfall, you'll notice that the lower frequency number (93.9500MHz) changes. This number tells us the frequency we will jump to if we click the mouse button, even if no station is there.

Also, notice that the frequency moves in jumps of 50kHz, the tuning step we specified when we configured this band profile. If we need to move in smaller steps, we can change the frequency step setting in Become a (Radio) Control Freak, on page 47.

Try This: Set a Private Bookmark

Having picked a frequency, we can bookmark it for later use. Let's try it. Click that bookmark icon, and a new dialog box pops up. The Frequency (in Hertz) is already filled in, as is the Modulation; we can change these if we like.

Give the bookmark a short Name, six characters or less, so that it will fit the resulting label.

Add a longer Description; this will appear when we hover the mouse over the bookmark.

Finally, tick the Scannable checkbox to add this bookmark to a list of stations to be scanned. This is best reserved for two-way radio channels, which come and go; broadcast stations will lock the scanner until the station goes off air.

Click OK, and the bookmark is set.

When you hover the mouse over a private bookmark, you can see the extended description you entered earlier; you can also edit or delete the bookmark.

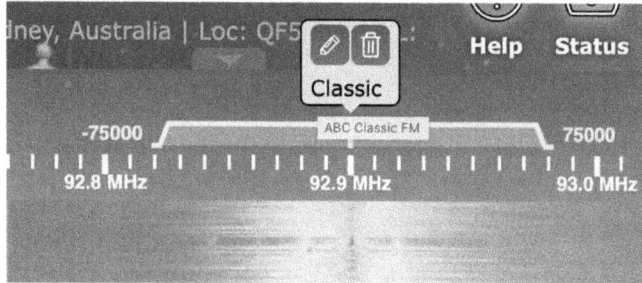

As an Administrator, we can also add bookmarks via Settings > Bookmark editor; those bookmarks are colored yellow and visible to all users. We'll come back to them in Shared Bookmarks, on page 53.

Demodulator Modes

On the broadcast bands, most stations use the same modulation type, usually AM or WFM, though we occasionally see other modes such as Synchronous AM, Digital Radio Mondiale (DRM), or Digital Audio Broadcasting (DAB).

For many other bands, we can expect a mixture of modulation modes, so it's useful to be able to switch demodulators as well.

Click the little triangle icon to expand the Modes section and show the available demodulators:

The most popular demodulators are on the top line:

- FM—Narrowband FM, typically used by two-way radio operators on VHF and UHF

- WFM—Wideband FM, used by VHF broadcast stations

- AM—Used by Medium Wave band broadcasters, some HF Citizens Band transmissions, and for Air Band communications

- LSB and USB—On HF, amateur radio operators typically use Lower Sideband below 10MHz, and Upper Sideband on 10MHz and above. When using digital modes, USB is the more common choice regardless of frequency. Non-amateur services on HF usually use USB regardless of the frequency. Some HF CB operators also use LSB or USB.

Try switching to some of the other demodulators to see the effect. In most cases, this will make it harder to understand the signal, though if you are listening to an AM signal, the LSB and USB demodulators will work fairly well, at least until you tune slightly off-frequency.

The buttons below these are for decoding data modes; we'll come back to them in Chapter 5, Explore with OpenWebRX+, on page 65.

Notice how the passband indicator changes when you switch between AM, LSB, and USB (we might need to zoom in to see the difference—we'll cover waterfall zoom controls in Display: Zooming About, Recording, and More, on page 51).

Become a (Radio) Control Freak

Many panel settings and controls let you really customize the OpenWebRX+ experience. You can mostly ignore them if you like (it's your receiver, after all), but some can be quite useful from time to time. Many settings have a corresponding "return to default" button, so if you find a change that you prefer, you can go to the Settings page and change the defaults permanently.

The Controls section is pretty straightforward, so the easiest way to become familiar with them is to try them and see what happens. If you hover your mouse pointer over a control, a pop-up will usually remind you what the control does. Some controls have more than one function, depending on which mouse button you click.

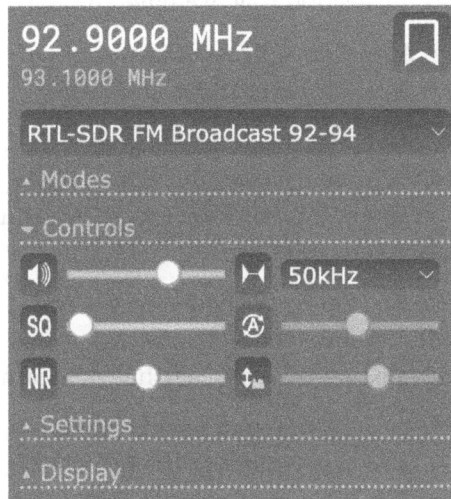

Try This: Volume and Frequency Step Controls

Let's tune in a station and use a few controls to see their effect.

On the first row, we have the following:

- Mute button
- Volume slider
- Tuning Step reset button
- Tuning Step drop-down list.

Click the Mute button to turn the audio on and off; when the audio is on, the volume slider works as you would expect, so we can adjust the sound to a comfortable level.

We've seen that on the FM broadcast band, mousing around the waterfall changes the displayed frequency in steps of 50kHz. Click the Tuning step drop-down and select a much smaller step. Now, when we mouse around, the displayed frequency changes are much finer.

Click the Tuning Step reset button to revert to the default tuning step we chose when we configured the band profile.

Try This: Squelch in the Signal Mud

On the second row, we have the Squelch controls:

- Auto Squelch/Scan button (SQ)
- Squelch Level slider

If you've ever used an FM walkie-talkie, you'll probably be familiar with the Squelch knob. When there is no signal present, you hear a lot of "white noise," and adjusting the Squelch control mutes the audio output, in this case, to spare your ears!

The Squelch Level slider serves the same purpose. The Auto squelch button sets the squelch level automatically, but we can use the slider instead to hear signals that are so weak that the software can't be sure they are real signals.

Try it out. Tune to a part of the band where there is no signal, and drag the Squelch Level slider to the left until you hear hissing audio; we say that "the squelch is open." Notice that the slider background is light green. Now drag the slider to the right until the light green turns dark. When you release the mouse button, the hissing stops, and "the squelch is closed."

Now, drag the slider to the left to reopen the squelch. Then, press the Auto squelch button, and the slider automatically returns to where the squelch closes. Any signal above the squelch level will be heard.

If you see a weak signal on the waterfall, click it. The Auto squelch may still prevent it from being heard, but we can manually drag the Squelch level to the left and hear the signal, though it may be noisy.

The Auto Squelch button has a second function, for scanning bookmarked stations, but we'll come back to that later.

Next to the Squelch controls are a couple of waterfall settings:

- Waterfall Color adjust button
- Waterfall Minimum Level slider

As we saw in The Art of Noise Reduction, on page 37, band conditions can vary, with different levels of background noise that can make it difficult to spot stations in the waterfall. OpenWebRX+ can deal with these changes automatically, but we have manual control as well if we want it.

The Waterfall Color adjust button sets the waterfall to give a good range of colors for the current band conditions; right-clicking the button tells the software to continually adjust things, which can be useful sometimes.

But we also have manual control of these levels: the Waterfall Minimum Level slider sets the signal level at which the waterfall shows dark blue.

The Waterfall Colors Reset button lets us hand control back to the software.

On the third row:

- Noise Reduction on/off button
- Noise Reduction Level slider
- Waterfall Colors Default button
- Waterfall Maximum Level slider

The noise reduction controls don't have much effect on broadcast FM, but they are very useful on HF to minimize various unwanted noise sources.

The Waterfall Colors Default button returns the minimum and maximum waterfall color levels back to their default settings.

The Waterfall Maximum Level slider is the companion to the Minimum Level slider in the second row; this control sets the signal level at which the waterfall shows red for the strongest signals.

Setting the Record (or the Waterfall) Straight

The Settings section of the receiver panel—not to be confused with the main OpenWebRX+ Settings pages—has more visual options related to the visual appearance of the receiver. If such things don't interest you, you can safely ignore most of this section, but at least have a look at the following checkbox settings:

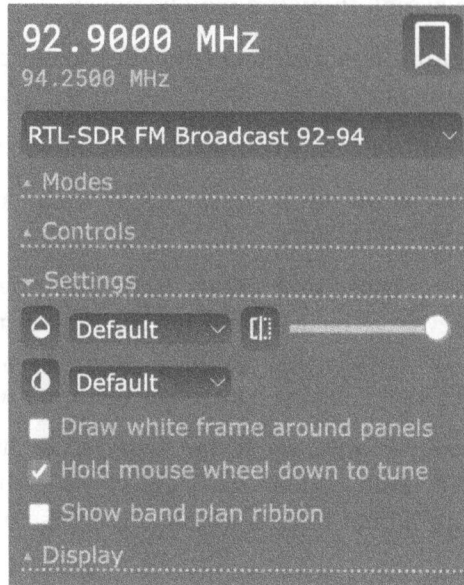

First row:

- Reset Panel Theme button—Returns the panel theme to the default setting
- Select Panel Theme—Try out different color schemes.
- Reset Panel Opacity button
- Set Panel Opacity slider—You might like to make your panels (semi-) transparent to let the waterfall shine through.

Second row:

- Reset waterfall theme to default
- Drop-down—Select a waterfall theme

Checkboxes:

- Draw white frame around panels (currently affects only the receiver controls panel)—This can make the panels stand out better on some themes.
- Hold mouse wheel down to tune—If ticked, then the mouse wheel zooms the display, similar to the KiwiSDR. If your mouse supports clicking with the mouse wheel, then holding the mouse wheel down while rolling it changes the receiver frequency. If the checkbox is unticked, the behavior is reversed.

- Show band plan ribbon—This is useful if our band profile covers several bands. Example: if we define a profile covering, say, 1.5MHz to 4MHz, this takes in the Medium Wave Broadcast, 160m Amateur, 120m Broadcast, 2MHz Aeronautical, 90m Broadcast, 3MHz Aeronautical, 80m and 75m Amateur, the upper 3MHz Aeronautical, and 75m Broadcast bands. It can be helpful to see the name of each band above the frequency gauge.

Display: Zooming About, Recording, and More

The most important parts in this section are the waterfall zoom controls: like the zoom lens on a camera, they let us get in to see the greatest level of detail or pull back to see the big picture.

First row:

- Zoom controls—In one level, out one level, zoom into the tuned signal, and zoom out to the full band

- Show/Hide spectrum display—Like the frequency domain graph we saw above the frequency gauge in Let's See What We Can Hear, on page 25

- REC (Record) on/off button—We can save the received audio to an MP3 file, so we can listen to it again later. The REC button "throbs" red while recording received audio; click it again to end recording. Mac users get prompted for the location of the saved audio file (defaulting to ~/Downloads), while for Windows users, the audio file always gets saved to the Downloads directory. In either case, the file is named REC-YYMMDD-HHMMSS-FFFFF.mp3, where YYMMDD-HHMMSS is the UTC date and time that the recording started, and FFFFF is the frequency in kilohertz. The recording file size is limited to 32MB, so it won't devour your hard drive if you forget about it.

Second row:

- The colored bar indicates relative signal strength—Normally green, turning to yellow and then red as the signal increases

Third row:

- Current UTC time—Universal Coordinated Time[10]
- Relative signal strength, as a number

---

10. https://www.timeanddate.com/worldclock/timezone/utc

Turn On, Tune In, but Don't Drop Out!

That was a lot of little details to absorb in a few pages; you don't need to remember everything, and most of the time, you will use only a few of them. Most are labelled with their function, and the rest will reveal a reminder if you hover your mouse over them.

Now that you have seen the main controls on the OpenWebRX+ user interface, you are probably also developing a good idea of what you would like this web SDR to cover. Whether it's shortwave broadcasting, aviation, two-way radio, amateur radio, or whatever, there is a lot of spectrum to explore!

If you like listening to the radio, you can probably stop here for a while, take in the sights (and sounds), and go deeper when and if you are ready for more.

Feel free to create new device/band profiles to explore any part of the spectrum that the receiver will cover; creating a profile takes less than a minute, and it can be deleted or modified with ease. Remember that your antenna should be able to work effectively on those frequencies, but try it and see.

Up Next: The Custom(iz)er is Always Right

At this point, you have a pretty good working OpenWebRX+ receiver, but there is still room for improvement. The generic configuration works pretty well, but you can make it suit your particular needs even better. You'll learn about regional band plans, bookmarking your favorite frequencies, and more. Ready? Let's get to it!

CHAPTER 4

Customize OpenWebRX+

We can now look at changes that will distinguish our SDR from all the others. Even if we don't make our receiver accessible to others on the web (we'll cover this in Chapter 6, Share Your Decoded Data, on page 77), we can make changes that suit our listening requirements best.

Shared Bookmarks

We saw in Try This: Set a Private Bookmark, on page 45 that we could create personal bookmarks via the receiver control panel. Those bookmarks are colored blue and are visible only on the browser where they were created because they are held in the browser's local storage area. We can turn any of these into shared bookmarks to make them visible to all users.

Try This: Convert Personal Bookmarks to Shared Bookmarks

Go to Settings > Bookmark editor > Import personal bookmarks; select any of your personal bookmarks from the list and click Import. The selected personal bookmarks are added to the shared bookmarks list. If you delete those bookmarks in the Bookmark editor, they revert to personal bookmarks.

We can also create bookmarks that are shared straight away. What kind of bookmarks might we create? Local broadcasters are a good place to start.

The EiBi shortwave database[1] that we've met before does not include local AM and FM broadcasters. Unlike shortwave, these stations generally reach only a few hundred kilometers at best, so they would only be of interest to listeners close to them. Many operate on the same frequency, as they are far enough apart not to interfere with each other.

---

1. http://eibispace.de/

Finding Broadcasters

Google "List of radio stations in (your country, state, city)" to find stations in your part of the world. Alternatively, Radio-Locator[2] can find nearby stations, especially in the United States.

Radio Garden[3] uses a Google Earth-style interface to find (and listen to) broadcasters in most parts of the world. However, some stations listed are often only streaming services, and they might not necessarily be transmitting radio signals. Still, it's a lot of fun to play with.

Armed with a list of nearby stations, we can bookmark some of them.

Try This: Bookmark Your Local Broadcasters

Let's pick a local station and create a new bookmark for it that will be visible to all users.

Go to Settings > Bookmark editor. You'll see the list of currently defined Bookmarks. Click "Add a new bookmark" at the bottom, and fill in the details for the station of interest:

- Name—JJJ-FM, a short nickname for the station
- Frequency—First select the units, MHz, from the drop-down, and enter the frequency, 105.7
- Modulation—WFM
- Underlying—None
- Description—ABC Triple-J FM
- Scan—Untick this checkbox, as scanning permanently stops on broadcast stations.

Finally, click the Save button on the right.

If we then return to the receiver window and select a band profile that includes the station, we'll see it has a bookmark label above it, this time in yellow. Click the bookmark label to tune in.

You can add more local stations in the same way. But using the Bookmark editor can be slow if you have a lot of bookmarks to add. We'll see a faster way in Try This: Bookmark Even More Stations, on page 63.

We'll come back to the Underlying setting in DIG a Little Deeper with Data Decoders, on page 68.

---

2.   https://radio-locator.com/
3.   https://radio.garden/

Other Things to Bookmark

You can bookmark non-broadcast stations in the same way. Here are a few examples to try:

- UHF CB—Out of the box, OpenWebRX+ includes HF (27MHz) CB channels, but not UHF CB channels. Like many short-range services, the details can vary a lot between regions, and there's no one set that suits everyone. The same applies to the Family Radio Service,[4] General Mobile Radio Service,[5] Multi-Use Radio Service,[6] and others in some regions.

- Air band frequencies, such as local airport towers, aviation and skydiving clubs, and so forth. A single large airport can have multiple channels in use, so it's largely up to the listener to set their own favorites.

- Ham radio clubs and other special interest groups have their own favorite channels, "net" frequencies, and so on, so these can be handy to bookmark as well.

Scanning Two-Way Radio Communications

All of these suggestions are two-way communications rather than broadcasts, so they are not continuous transmissions. So, ticking the Scannable checkbox lets us monitor many channels rather than just one. To scan them, right-click the SQ button in the receiver Controls section.

Need more ideas? Check out this blog post[7] about the VK2MB KiwiSDR to see how my club customized it.

Check Under the Hood

While we can customize a lot of OpenWebRX+ using its web interface, it can be a lot quicker to access the underlying files using the Secure Shell (SSH). With it, we can use the host's command line remotely from our local computer.

This should only be attempted if you are comfortable with the Linux file system and working with the Linux command line; it's also a good idea to use a version control system such as Git[8] to manage versions of each file you modify.

4.　https://www.fcc.gov/wireless/bureau-divisions/mobility-division/family-radio-service-frs

5.　https://www.fcc.gov/wireless/bureau-divisions/mobility-division/general-mobile-radio-service-gmrs

6.　https://www.fcc.gov/wireless/bureau-divisions/mobility-division/multi-use-radio-service-murs

7.　https://www.mwrs.org.au/2024/03/05/the-kiwi-has-landed/

8.　https://git-scm.com/

Lord of the Files

Many of the configuration files live in the /var/lib/openwebrx/ directory. Most are best managed using the Settings pages, but /var/lib/openwebrx/bookmarks.json can be worth editing if we have a lot of bookmarks to manage.

JSON Format

Take care with punctuation: the JSON[9] format these files use is simple but very strict: one missing or misplaced comma or a mismatched brace can render the file unusable, or at least a pain to correct.

jq[10] is a useful tool for finding formatting problems in a JSON file. It's easy to install. SSH into the host, and install like this:

```
$ sudo apt install jq
```
Note: sudo is short for "Super User Do," or "Run as Administrator" as they say in Microsoft circles.

Then you can check any JSON file, say, TEST.json:

```
$ jq . TEST.json    # (note the dot between jq and the file name)
parse error: Expected separator between values at line 7, column 0
```
We could then edit the file and go straight to line 7:

```
$ nano -l TEST.json    # ("-l" enables line numbers in nano)
1 {
2   "items": [           # start of array
3     "computer",
4     "RTL-SDR dongle"    # comma missing at the end of line
5     "antenna",
6     "power supply"      # last item doesn't need a comma
7   ]                     # error reported at the end of the array
8 }                       # (note: comments are not valid in JSON)
```
Sometimes, the actual error is on the preceding line, or a bit earlier, but jq gets us close.

In this example, the error is reported as line 7, the end of the items array. But the actual error is on line 4, in the middle of the array, because we're missing the comma separator between RTL-SDR dongle and antenna.

---

9.   https://www.json.org/json-en.html

10. https://jqlang.org/

Before making any file changes, we should shut down OpenWebRX+, or we risk having our changes overwritten by the OpenWebRX+ server. SSH into the OpenWebRX+ server and use systemctl:

```
$ sudo systemctl stop openwebrx
```

After we have completed our changes, we will restart OpenWebRX+:

```
$ sudo systemctl start openwebrx
```

Bookmark Files

Now, to the files we can edit:

- /var/lib/openwebrx/bookmarks.json—As the name implies, this file contains any bookmarks that we have set with the Bookmark editor; if we wish to set a large number of bookmarks, editing this file can be a lot quicker than using the web interface.

- /etc/openwebrx/bookmarks.d/—This subdirectory contains a number of JSON files with spot frequencies of interest for particular radio services:

  - aviation.json—Aircraft communication frequencies, for both voice and data communications; selecting a data mode will open the corresponding decoder window. The file does not contain local airports or non-directional beacon (NDB) frequencies, so these can be worth adding.

  - cb.json—27MHz HF CB channels; no UHF channels are included, as these vary between regions, but we can add them if we like.

  - gmrs.json—General Mobile Radio Service[11] channels (United States and Canada only), low-power UHF two-way radio communications. The channels are shared with the lower-power Family Radio Service.

  - maritime.json—Services used by ships: NAVTEX and MSI (both of which we can view using the SITOR-B decoder), DSC (Digital Selective Calling), and so on

  - misc.json—Time signals, markers, assorted noises like Over the Horizon Radar (OTHR), known jammers, and so on

  - murs.json—Multi-Use Radio Service,[12] for unlicensed two-way voice and data communications in the United States

---

11. https://www.radioddity.com/blogs/all/everything-you-want-to-know-about-gmrs-radio
12. https://www.fcc.gov/wireless/bureau-divisions/mobility-division/multi-use-radio-service-murs

- pmr.json—Private Mobile Radio (PMR466), similar to GMRS, in the European Union

- weather.json—Weather-related services such as HF RTTY

- wfax.json—Weather facsimile stations

Three folders, r1, r2, and r3, contain spot frequency files specific to each of the three ITU regions, for example, r2/noaa.json has the NOAA VHF weather channels in the United States (Region 2). Other folders, named for their two-letter country codes, contain frequency information for services specific to those countries.

We can also edit the regional band plan files, as we will see shortly, in Strike Up the Band Plan, on page 59.

Hands Off, Please

Some files are best left alone, because OpenWebRX+ periodically updates them from online sources, overwriting any local changes we might have made:

- /var/lib/openwebrx/eibi.json—Shortwave broadcast stations, updated daily from eibis-pace.de

- /var/lib/openwebrx/markers.json—Map locations of broadcast stations, repeaters, weather stations, and online receivers, viewable on the Map page

- /var/lib/openwebrx/repeaters.json—Amateur radio repeaters within the range we specified in Settings > General Settings > Display settings; updated daily from RepeaterBook

- /etc/openwebrx/openwebrx.conf.d/20-temporary-directory.conf—Holds the location of OpenWe-bRX+'s temporary working directory. There should be no need to change this.

There are a couple of other files that we can more easily manage from the Settings web interface:

- receiver_top_photo.png—The receiver panorama photo we specified in Settings > General Settings > Receiver images

- settings.json—Assorted configuration values

Try This: Add Local Air Band Frequencies

In your web browser, search for "Airport frequencies near me." This should bring up a list of your local air band channels, or try the website 16 Right.[13]

---

13. https://16right.com/spotting/frequencies

The Radio Reference Wiki's Aircraft[14] and Finding Air Traffic Frequencies[15] pages have plenty of suggestions.

And, Frugal Radio's Monitoring Aviation Communications playlist[16] has some great video tutorials.

Here in northern Sydney, 16 Right includes "Departures North & East" on 123.0, and "Approach - North" on 124.4. These two frequencies will fit in a single RTL-SDR band profile, as they are less than 2MHz apart.

Create a band profile named (in this example) "Air 123-125", centered on 124.0MHz, sampling at 2.5 MS/s, with an initial frequency of 124.4MHz, a default modulation of AM, and a 10kHz step size. Refer back to Try This: Create a Band Profile, on page 22 for a refresher.

We can add bookmarks for the 123.0 and 124.4 using the Bookmarks editor, of course, but we can also do it by editing /etc/openwebrx/bookmarks.d/aviation.json.

SSH or PuTTy into the OpenWebRX+ server, and enter these commands:

```
$ sudo systemctl stop openwebrx
$ cd /etc/openwebrx/bookmarks.d/
$ cp aviation.json aviation-backup.json
```

Open the file with your favorite text editor, for example, sudo nano aviation.json. Save and exit the changes, and restart OpenWebRX+:

```
$ sudo systemctl start openwebrx
$ exit
```

Go back to OpenWebRx+ and open the new band profile ("Air 123-125"). The two frequencies are now labelled for easy selection.

Strike Up the Band Plan

A third kind of bookmark, determined by the official regional band plans, is typically used on the amateur radio bands. These bookmarks show frequencies where it's common practice to use a particular communication mode.

On the RTL-SDR 2m band profile, we can see green bookmarks for modes like JT65, FT8, WSPR, PACKET, and others; these are data modes where particular software is used to create the required signals. By convention, these frequencies exist so that people using these modes can communicate with each other.

---

14. https://wiki.radioreference.com/index.php/Aircraft
15. https://wiki.radioreference.com/index.php/Finding_Air_Traffic_Frequencies
16. https://www.youtube.com/playlist?list=PLe5ZKeM2hRBlLfL6peRIcETng_DLrm5Kt

Any station could appear on these spot frequencies using these data modes, and if you click the marker, OpenWebRX+ will open a suitable decoder.

Think Regional, Act Local

The International Amateur Radio Union[17] divides the world into three regions:

- Region 1—Africa, Europe, Middle East, Northern Asia[18]
- Region 2—The Americas[19]
- Region 3—Southern Asia, Australia, Oceania[20]

The name of the corresponding file for each region includes the region number, for example, the Region 3 band plan file is /etc/openwebrx/bands-r3.json. If you're in Region 1 or 2, the filename would use the corresponding region number.

The band plan file specifies the various amateur radio band limits and the recommended modes to use on particular frequencies in each band. Because IARU regions are large, band plan differences exist between countries in the same region, so it can be useful to edit our regional band plan file to suit our country.

For example, at the time of writing, the Region 3 band plan file /etc/openwebrx/bands-r3.json shows 144.800MHz as a packet radio frequency. But that's not true everywhere in Region 3. In Australia, that frequency is assigned as a digital narrow band calling frequency.

Try This: Update the Band Plan File

The *Automatic Packet Reporting System* (APRS)[21] frequency is 145.175MHz in Australia, 144.575 in New Zealand, 144.39 in Indonesia, 145.525 in Thailand, and 144.64 in China. All these countries are in Region 3, so it's a good idea to adjust the band plan file to reflect your local conditions; similar changes may be required in Regions 1 and 2.

Around the World, the Trip Begins with the ISS

While we're here, we'll add an SSTV spot frequency, because the International Space Station sometimes transmits pictures on 145.800MHz FM. ISS uses this frequency in all IARU regions.

---

17. https://www.iaru.org/
18. https://www.iaru-r1.org/
19. https://www.iaru-r2.org/en/
20. https://www.iaru-r3.org/
21. https://www.areg.org.au/activities-old/automatic-position-reporting-system-aprs

SSH into the OpenWebRX+ host, and use the Linux text editor of your choice to open the band plan file. If you're new to Linux and want to keep things simple, the "nano" or "pico" editors[22] will serve you well.

```
$ sudo systemctl stop openwebrx
$ sudo nano /etc/openwebrx/bands-r3.json
```

Note the sudo in the command: we need to use administrator privileges to make changes in the /etc directory.

As a guide, after my changes for Australia, the two-meter band section of the file looks like this:

```
{
  "name": "2m",
  "lower_bound": 144000000,
  "upper_bound": 148000000,
  "frequencies": {
    "wspr": 144489000,
    "ft8": 144174000,
    "ft4": 144170000,
    "jt65": 144120000,
    "packet": [145175000, 145825000], // was [144800000, 145825000]
    "q65": 144116000,
    "msk144": 144360000,  // changed: added a comma at the end
    "sstv": { "frequency": 145800000, "underlying": "fm" } // new
  },
  "tags": ["hamradio"]
},
```

> Note that the // was..., // changed..., and // new comments are not part of the JSON file format. They will cause an error if we leave them in there.

If you installed the jq utility, run a quick check on the file we just modified:

```
$ jq . /etc/openwebrx/bands-r3.json
```

If jq reports any errors, it's best to correct them straight away.

We can then start OpenWebRX+ again:

```
$ sudo systemctl start openwebrx
```

Refresh the OpenWebRX+ receiver tab in the browser, select the existing 2m band profile; at 145.175, you should see the new PACKET marker, and at 145.800, you should see SSTV.

---

22. https://www.computerhope.com/unix/upico.htm

Other bands in the file have the same structure, so we can change them in the same way if required.

What Lies Beneath...

In the "sstv" section, the "underlying": "fm" part is worthy of comment. SSTV images are transmitted as audio tones so that they can "fit" inside a normal radio voice channel. The underlying modulation method is usually the same as the voice mode in the given band. On HF bands below 10MHz, SSTV signals are sent as LSB; on 10MHz and above, they are USB.

We need to include this "underlying" modulation mode with the frequency, so instead of writing this:

```
"sstv": 145800000
```

... we wrote this:

```
"sstv": { "frequency": 145800000, "underlying": "fm" }
```

According to the Amateur Radio on the ISS (ARISS) status page,[23] SSTV transmissions from the Space Station use a VHF FM transceiver, so "underlying" is set to "fm". Incidentally, the 145825000 in the "packet" field is also used by the ISS, so we can track it with APRS as it flies over us.

The frequencies for the other modes (WSPR, FT8, and so on) might also be incorrect for any given country; it's worth checking band plans published by your local IARU representative organization. The Signal Identification Guide also has frequency lists for many digital modes.

Radio at Your Service

We saw in Bookmark Files, on page 57 that we have several files for holding frequency bookmarks, each for different services like aviation, maritime, and so on. Let's take a closer look.

The service bookmark files are kept in the /etc/openwebrx/bookmarks.d directory. They all have a similar JSON format, with each entry having three parts:

- A concise name for the displayed marker
- The frequency
- The modulation, which tells us which demodulator to use

---

23. https://www.ariss.org/current-status-of-iss-stations.html

For example, here's an entry in the weather facsimile stations file, /etc/openwe-brx/bookmarks.d/wfax.json. The entry is for the Australian Bureau of Meteorology's VMC[24] station in Charleville, Queensland, on 5100kHz:

```
{
    "name" : "VMC - Charleville",
    "frequency" : 5098100,
    "modulation" : "fax"
}
```

Notice that the frequency is 5098100 (Hz), or 5098.1kHz, instead of the official 5100kHz. It's common to see slight frequency offsets like this "behind the scenes" to help tweak the performance of the OpenWebRX+ fax decoder. We don't need to worry about it since we click the label to tune the station.

Try This: Bookmark Even More Stations

Our shared bookmarks file, /var/lib/openwebrx/bookmarks.json, which is managed by Settings > Bookmarks editor, has a similar format. Here's the entry for ABC Classic FM that we saw earlier:

```
{
  "name": "Classic",
  "frequency": 92900000,
  "modulation": "wfm",
  "description": "ABC Classic FM",
  "scannable": false
}
```

Note that we have two extra parts: 1) a more verbose description that pops up when we hover our mouse over the marker, and 2) a "scannable" checkbox.

Add some more of your favorite stations. Go to the end of the file, add a comma, and then another entry in the same format; for example:

```
{
  "name": "ABC Sydney",
  "frequency": 702000,
  "modulation": "am",
  "description": "ABC Local Radio Sydney",
  "scannable": false
}
```

Add as many of your local stations as you like, on any frequency you like. Remember to put a comma between each one. The very last entry in the file should not have a comma at the end.

---

24. http://www.bom.gov.au/marine/radio-sat/vmc-technical-guide.shtml

The order of these entries is not important, as the software will sort them out.

After you have added your station bookmarks, close the file, restart OpenWe-bRX+, select the relevant band profile, and the bookmarks should now appear.

Up Next: Let's See What We Can't Hear

By now, your OpenWebRX+ receiver is more "at home," with bookmarks for favorite stations and other signals of interest. As you continue to learn, you can always come back and update your configuration, add more bookmarks, create new band profiles, and more.

Next, we'll dive into some of the more exotic features of the software: decoding data transmissions, reporting the received data to online aggregators, mapping aircraft and other stations, and customizing OpenWebRX+ to make it truly personal. Onward!

CHAPTER 5

Explore with OpenWebRX+

In Demodulator Modes, on page 46, we saw that we can do more with Open-WebRX+ than listen to voices and music. There are more exotic sounds to be heard on the radio, even if we can't decipher them by ear. In this chapter, we'll explore some of the less common voice modes and the hidden world of data modes.

Demodulator Modes, Revisited

Let's take another look at the Receiver control panel. Below the FM, WFM, and other voice mode buttons, there are three more rows of demodulators.

CW (Continuous Wave, or Morse Code) is the granddaddy of digital communication modes, which dates back to the wired telegraph days of the early 19th century. It's also a mode born of tragedy.[1] We'll come back to CW presently.

```
92.9000 MHz              🔖
94.2500 MHz

RTL-SDR FM Broadcast 92-94    ⌄

▾ Modes

FM   WFM   AM   LSB   USB
CW   SAM   DATA   M17
FreeDV   DRM   DAB   HDR
DIG                       ⌄

▸ Controls

▸ Settings

▸ Display
```

Other Demodulators

A number of other demodulators are available here. They're outside the scope of this book, but don't let that stop you from exploring! Check out these modes on the Signal Identification Guide; see the footnotes for the modes that appeal to you. They are described briefly on the next page.

---

1. https://vocal.media/fyi/the-tragic-story-behind-the-morse-code

SAM or Synchronous Amplitude Modulation

is a variation on normal AM broadcasting that can improve reception under certain circumstances. If your AM signal is fading in and out, try switching to SAM. You can find out more about it on the Radio Reference Wiki.[2]

DATA

is not a specific communications mode; selecting this makes the receiver's IQ data stream available to an external tool such as CSDR,[3] which opens up many avenues for experimenting.

M17

is a fairly new open source digital radio mode for amateur radio applications; check it out on the M17 Project[4] website.

FreeDV[5]

is an open source HF Digital Voice mode. Analog voice transmissions on the HF bands can sometimes be difficult to listen to, suffering noisy interference from static crashes, fading, and so on, but FreeDV allows audio quality comparable to FM. There is even an enhancement, FreeDV+,[6] which includes video, but OpenWebRX+ does not currently support it.

DRM (Digital Radio Mondiale)[7]

is a digital broadcasting standard providing FM-quality audio on the HF bands. Check the DRM Reception Project[8] for broadcast schedules. The KiwiSDR's DRM decoder also makes it easy to check these transmissions.

DAB (Digital Audio Broadcasting)

is slowly edging out wideband FM for high-quality audio, but it still has a way to go. The original DAB is being phased out in favor of DAB+. Check DAB Ensembles Worldwide[9] for frequencies in your area.

HDR (High Data Rate)[10]

are modem signals favored by some military operators.

I promised we'd come back to CW, so here we are.

---

2. https://wiki.radioreference.com/index.php/Synchronous_Detection
3. https://github.com/jketterl/csdr
4. https://m17project.org/
5. https://freedv.org/
6. https://www.sigidwiki.com/wiki/FreeDV_plus_Video
7. https://www.sigidwiki.com/wiki/DRM
8. https://www.drmrx.org/
9. https://www.wohnort.org/dab/
10. https://www.sigidwiki.com/wiki/CIS_OFDM_HDR_Modem

Try This: You Too Can See CW

These days, CW is still very popular among amateur radio operators. With practice, it's not too hard to decode most CW signals by ear. There are phone apps, computer programs, and courses[11] to help you learn "The Code."

OpenWebRX+ also has built-in CW Decoder and CW Skimmer extensions, which can help us to follow conversations until we can read Morse Code by ear. A warning, though: most Morse decoders perform well only on strong signals, without much noise on the band.

The first thing we need to do is find a suitable Morse signal on the waterfall. These are typically found at the low-frequency end of each amateur band, and on some aviation beacons on VLF.

The defining characteristic of Morse is an audio tone pulsing on and off to form "dots" and "dashes" (or "dits" and "dahs" as the radio hams prefer to call them, because that's how they sound).

To get a feel for the sound of Morse Code, listen to the VK2WI Morse Practice beacon[12] on the VK2MB KiwiSDR in Sydney. The decoded Morse should also appear above the waterfall; if it doesn't, click the VK2WI bookmark.

Back on your SDR, click a Morse signal in the waterfall. Then press the CW button to narrow the passband and filter out any other signals that might be nearby, so you can focus on that one Morse signal.

You can also select "CW decoder" from the drop-down list to the right of the DIG button. This opens a mini waterfall and decoder text pane at the bottom left of the screen. Clicking again on DIG closes the decoder.

When we hover over the mini waterfall, a yellow vertical bar appears. Click the Morse signal in the mini waterfall to position the yellow bar on top of the signal. With a bit of luck, the decoded Morse text should start to appear in the text pane. This is shown in the image at the top of the next page.

There are a few things to notice about this screenshot. OpenWebRX+ was monitoring a CW test signal (VVV DE VK2SKY) in an environment with a fairly high level of electrical noise. To a CW receiver, such noise often looks like short bursts of carrier, and so can be confused with the wanted signal, dropping many E (dit) and T (dah) characters in the decode window.

---

11. https://courses.wavetalkers.com/p/learn-morse-code-the-basics
12. http://websdr.mwrs.org.au:8073/?f=3699.00cwz10

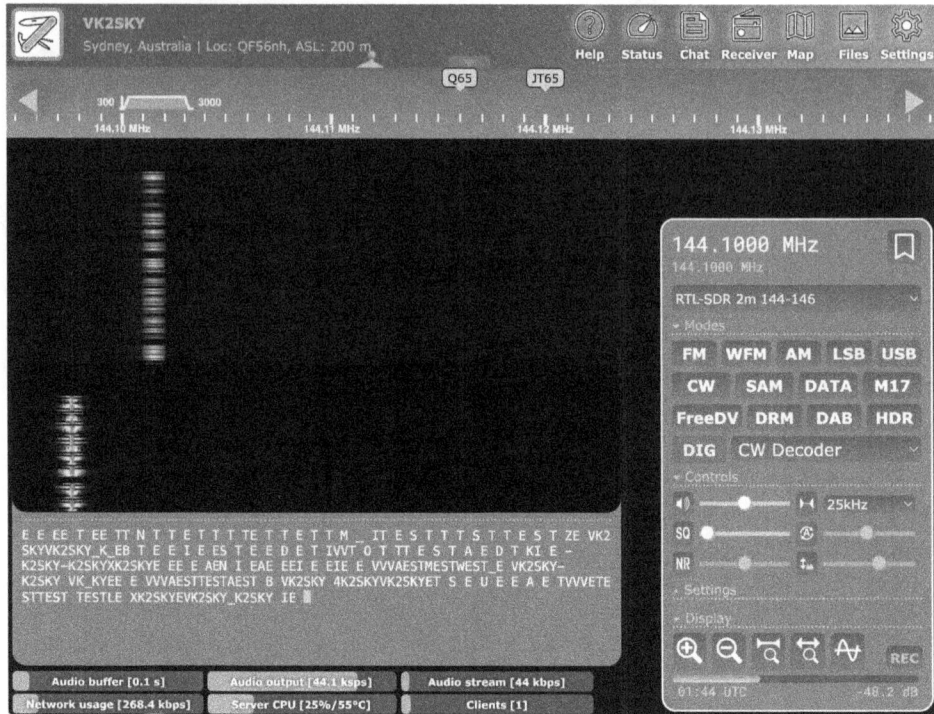

The Kiwi's CW decoder has a threshold setting that lets the user filter out a lot of the atmospheric noise, and it can adapt to varying speeds. The Open-WebRX+ CW decoder doesn't support these features yet.

The CW Skimmer,[13] below the CW Decoder in the drop-down list, can decode multiple Morse signals at the same time. This can be handy during Morse contests when the bands are crowded with many CW stations vying for the greatest number of contacts.

DIG a Little Deeper with Data Decoders

OpenWebRX+ offers many different data decoders, and new decoders seem to appear every few versions, so we'll try out only the more popular ones for now.

---

13. https://github.com/luarvique/csdr-cwskimmer

The Underlying Modulation

Back in What Lies Beneath…, on page 62, when we were customizing our band plan file to deal with SSTV transmissions from the International Space Station, we came across the notion of an "underlying modulation" of the signal. On the station, SSTV pictures are converted to an audio stream and fed to an FM transmitter, so FM is the underlying modulation.

This idea pops up with other digital modes. For example, selecting ACARS, the Aircraft Communications Addressing and Reporting System[14] decoder, also highlights the AM mode button because ACARS data is converted to sound and transmitted using the underlying AM modulation for compatibility with Air Band voice transmissions.

With SSTV mode, the FM, LSB, and USB demodulator buttons all get highlighted in yellow because any of them could underlie the picture's audio stream. We need to select one of them, depending on the band we transmit on:

- LSB for amateur radio on HF frequencies below 10MHz
- USB for amateur radio HF frequencies above 10MHz, or non-amateur transmissions anywhere on HF
- FM for VHF transmissions, such as on the ISS

Try This: Decode ACARS Status Messages

Air travel is such a part of modern life that in most parts of the world, airliners pass over our heads every hour of every day. If in doubt, visit Flightradar24[15] or one of the other popular flight tracking websites.

The pilots and air traffic control talk a lot, and we can monitor those transmissions, but the aircraft themselves have a lot to say, too! Let's create a band profile to capture the data that they transmit. Refer back to Try This: Create a Band Profile, on page 22 if you'd like a refresher.

We'll use these settings:

- Profile name—ACARS test
- Center frequency—131.55MHz, the World Primary ACARS channel
- Sample rate—2.5 MS/s

---

14. https://www.sigidwiki.com/wiki/ACARS
15. https://www.flightradar24.com/

- Initial frequency—131.55MHz
- Initial modulation—ACARS
- Tuning step—10000 Hz

Click "Apply and save" and then go back to the receiver tab and select "RTL-SDR ACARS test" from the band profile drop-down. Then, wait for aircraft to fly overhead. After a while, ACARS messages should start appearing in a panel on the left side of the waterfall, like this:

Time	Flight	Aircraft	Data	Clear
22.710.268 96.81 96.812711-00001100 523				
11:34:28	QF749	VH-VXA	ACARS frame	
1444 94.47 94.063736++++++"1 20.8 25.810.286 98.72 98.815711-00001100 1677 1588 94.53 94.123736++++ ++"1 20.5 25.510.295 98.72 98.810211-00001100 1805 1715 94.59 94 .163736++++++"1 20.0 2				
11:34:32	QF749	VH-VXA	ACARS frame	
5.210.305 98.72 98.810711-00001100 1963 18 72 94.59 94.163738++++++"1 19.8 25.210.311 98.72 98.751211-0000110 0 2113 2024 94.56 94.123737++++++"1 19.3 25.010.320 98.66 98.691 711-00001100 2209				
11:34:36	QF749	VH-VXA	ACARS frame	
2119 94.38 93.943636++++++"1 18.8 25.210.332 9 8.59 98.632211-00001100 2304 2215 94.25 93.813636++++++"1 18.5 2 5.210.343 98.53 98.562711-00001100 2471 2380 94.16 93.753636++++++' '1 18.0 2				
11:34:43	QF749	VH-VXA	ACARS frame	
2915 94.03 93.633636++++++"1 17.0 25.010.372 98.41 98.47				

Notice that in the third column, the aircraft registration, VH-VXA, is underlined; it's a browser link. We can click it to see details of the aircraft and its current flight on the FlightAware[16] tracking site.

The rest of the ACARS data can be very cryptic, though plain text messages can often be seen, too. Here are a couple more resources to help you understand this mode:

- ACARS Introduction[17]
- ACARS on the Signal Identification Guide[18]

---

16. https://www.flightaware.com/
17. https://www.universal-radio.com/catalog/decoders/acarsweb.pdf
18. https://www.sigidwiki.com/wiki/ACARS

Try This: Track Aircraft Movements with ADS-B[19]

We can also follow the movements of the aircraft, often seeing flights that get filtered out of the commercial flight tracking websites.

Let's create another profile, called ADS-B Test, with a center frequency of 1090MHz and an initial modulation of "ADSB." With this new profile active, a table builds up on screen with ADS-B data transmitted by nearby aircraft:

Flight	Aircraft	Squawk	Dist	Alt (ft)		Speed (kt)		Signal
UAL839	N27957	7305	7 km		9375	SSE	280	−29 dB
ACA33	C−FIUA	2776	8 km	64↑	9425	SSE	266	−29.6 dB
	7C6B0C	1442	16 km	2176↑	10800	WNW	317	−27.8 dB
JST401	7C6C92		9 km	384↓	6300	S	300	−28.8 dB

This table contains different information than ACARS mode:

- Flight—The flight number may or may not appear; if it does, click it for more flight information.
- Aircraft—The ICAO (International Civil Aviation Organization) 24-bit identifier;[20] this field can sometimes show the aircraft tail number instead. In either case, click it for more aircraft information.
- Squawk—The aircraft's Squawk code, a four-digit numeric code assigned by local air traffic controllers to tag the aircraft on their radar screens. Codes 7500, 7600, and 7700 signify onboard emergencies.[21]
- Dist—The aircraft's distance from your receiver; click to show the aircraft on a map.
- Alt (ft)—Ascent or descent rate and altitude, both in feet
- Speed (kt)—Aircraft heading and speed in knots
- Signal—The relative signal strength

The Map display, shown on the next page, opens in a new browser tab so we can see the movements of all detected aircraft on a display similar to commercial tracking websites.

Later, in Visit the Linking Memorial, on page 74, you'll learn how to control what happens when you click the links in this table.

---

19. https://www.sigidwiki.com/wiki/Automatic_Dependent_Surveillance-Broadcast_(ADS-B)
20. https://www.aerotransport.org/html/ICAO_hex_decode.html
21. https://www.youtube.com/watch?v=3BpIPDYHHls

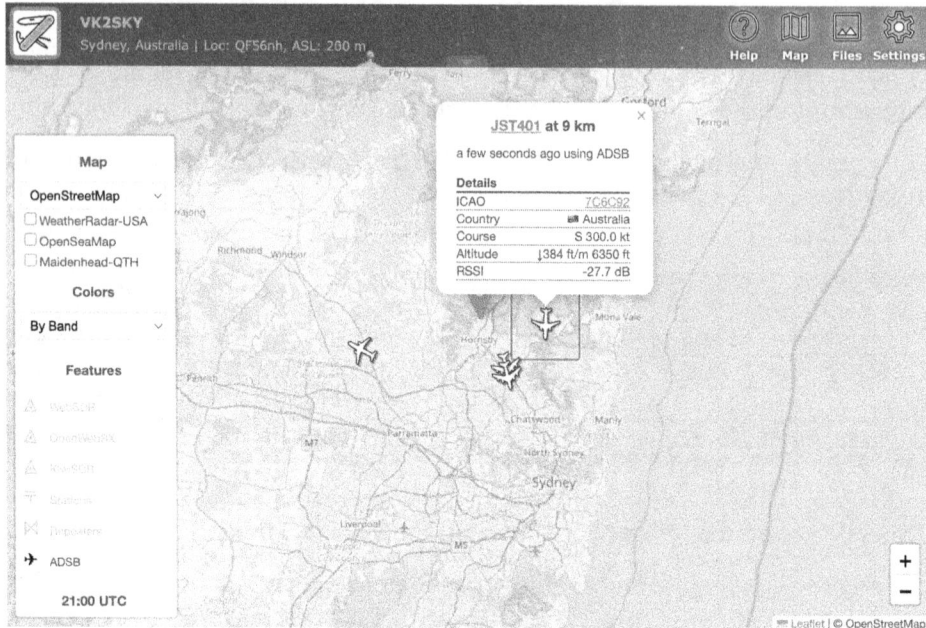

Try This: Track Vehicles with APRS

The Automatic Packet Reporting System[22] is a tactical communications pro-tocol over packet radio.[23] One popular use of APRS is to send vehicle positions, so we'll sometimes hear it referred to (incorrectly) as the Automatic *Position Reporting System*.

But APRS is also useful for communicating other situational information like weather station data, and hazards such as fires, floods, and earthquakes. The popular website, APRS.fi,[24] displays information gathered by the APRS Internet Service (APRS-IS) from all over the world.

We can select a known packet radio frequency, such as 145.175MHz, the Australian national APRS channel. If we checked and updated our regional band plan file in Think Regional, Act Local, on page 60, then the frequency will have a green PACKET bookmark. We can click it to open the packet/APRS decoder panel to monitor the traffic.

If there is no bookmark, simply tune to the frequency and select Packet from the Digital Modes drop-down list.

---

22. https://www.sigidwiki.com/wiki/APRS
23. https://www.sigidwiki.com/wiki/PACKET
24. https://aprs.fi/

In either case, a mini-waterfall and decoded data panel should pop up on the left side of the screen.

UTC	Callsign	Coord	Comment	Clear
	VK2FJ-9		Listening to DMR 505	
	VK2BWI-1		VK2WI Amateur Radio NSW QF56MH	
	VK2WIV-5		WICEN VRA-62 Batt:14.2V Temp:29C	
	VK2BWI-1			
	VK2WIV-5		WICEN VRA-62 Batt:14.1V Temp:30C	
	VK2BWI-1			
	VK2BWI-1		VK2WI Amateur Radio NSW QF56MH	

Notice that the call signs in the second column are clickable links; these take us to a website with more info about the clicked station.

In What's So Great About Aggregators?, on page 77 we'll see how to share this received information with the APRS Internet Service.

Other Digital Decoders to Try

OpenWebRX+ supports many other digital decoders, which we can pick from the adjacent drop-down list. What each of them is used for is outside the scope of this book, but searching for the modes on the Signal Identification Guide will get you started with frequencies and identifying signals by ear.

These are some of the better-known modes in the list:

AIS
    Automatic Identification System, like ACARS and ADS-B, but for tracking maritime vessels

Fax
    Facsimile images, typically weather service charts

FT4 and FT8
    Popular in the amateur radio community, these "weak signal" modes can be effective for very long distance communications at very low power levels; received signal reports are collected online by the PSK Reporter[25] aggregator.

RTTY
    Radioteletype is another data communications mode popular with amateurs and news agencies. There are several RTTY selections on the list and common presets for different speeds and data formats.

---

25. https://pskreporter.info/pskmap.html

SSTV

As previously discussed, the International Space Station sometimes transmits SSTV images, as do many hams on the ground. Received images are archived and can be accessed by clicking the Files button at the top of the screen.

WSPR

Weak Signal Propagation Reporter, pronounced "whisper," is another popular mode. Reports of received WSPR signals also go to PSK Reporter. Some researchers believe that analysis of archived WSPR signal data could help pinpoint the final resting place of Malaysian Airlines flight MH-370,[26] but this has yet to be proven.

Visit the Linking Memorial

We've already seen that OpenWebRX+ displays some information as clickable links—things like radio call signs, airline flight numbers, and aircraft identification codes. When we click these links, we jump to external websites that give us more information.

We can also decide which websites to link to for that information. In Settings > General Settings, we can scroll down to External links, where we'll find these particular settings:

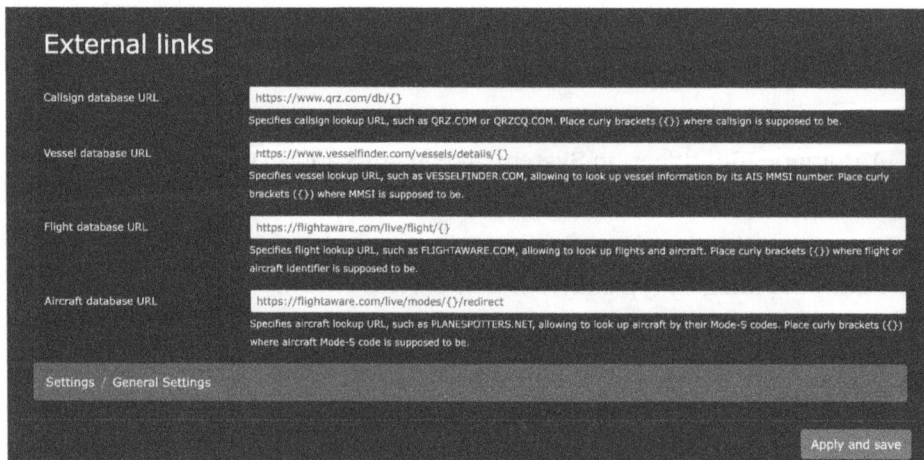

This section has the URLs for the external databases. You can leave them at their defaults, or change them if you like. Here are a few changes to consider.

---

26. https://www.mh370search.com/category/wspr/

Try This: Change the Amateur Call Sign Database

The "Callsign database URL" determines where we'll go if we click a call sign field, as we did in Try This: Track Vehicles with APRS, on page 72.

Out of the box, this field is set to refer to the QRZCQ website:

```
https://www.qrzcq.com/call/{}
```

Notice the "{}" in the URL, which is a placeholder for the call sign.

QRZCQ is a directory of amateur radio call signs; a more popular site is QRZ. (QRZ is the ham radio code for "Who is Calling Me?")

For an admittedly biased view, compare the entries for VK2VMP, onboard HMAS Vampire, at the Australian National Maritime Museum in Sydney: qrzcq.com[27] and qrz.com[28].

If you prefer the second page, change "Callsign database URL" to "'

```
https://www.qrz.com/db/{}
```

Any website URL can work here, as long as it expects a radio call sign. Include the call sign placeholder, "{}", in the right part of the URL.

Try This: Change the Vessel Database

The Automatic Identification System decoder also uses an external database of marine vessels, using its Maritime Mobile Service Identity (MMSI).

The default value for this field is Vessel Finder:

```
https://www.vesselfinder.com/vessels/details/{}
```

Again, notice the "{}" placeholder, which in this case is for the MMSI.

Two alternative values you can try here:

Marine Traffic
```
https://www.marinetraffic.com/en/ais/details/ships/mmsi:{}
```

My Ship Tracking
```
https://www.myshiptracking.com/?mmsi={}
```

Pick whichever you prefer, or you can go back to Vessel Finder.

---

27. https://www.qrzcq.com/call/VK2VMP
28. https://www.qrz.com/db/VK2VMP

Try This: Change the Flight Database

We can look up an aircraft by a combination of an airline code and the aircraft's flight number.

Out of the box, "Flight database URL" is set to FlightAware: https://www.flightaware.com/live/flight/{}

You can also try Flightradar24:

```
https://www.flightradar24.com/{}
```

Try This: Change the Aircraft Database

Finally, we can track aircraft by their ICAO 24-bit address. The default URL for this is also on Flight Aware:

```
https://flightaware.com/live/modes/{}
```

ADS-B Exchange is another good option.

```
https://globe.adsbexchange.com/?icao={}
```

With all of the external links, new services may appear, and as long as a service accepts the right piece of identifier in the URL, we can use it, remembering to include the "{}" placeholder.

Up Next: Sharing is Caring

So, now you can decode the digital signals you receive and fetch more information about some of those signals using various online services.

But wait, there's more. You can take your decoded data and feed it to online aggregators so that others can make use of it.

You'll also learn more about the OpenWebRX+ receiver directory and how to share your receiver online. Let's go!

CHAPTER 6

Share Your Decoded Data

In Chapter 5, Explore with OpenWebRX+, on page 65, we were able to receive data signals in many forms and display the decoded data on our screens. OpenWebRX+ can also take much of that data and share it with the world using online aggregators.

What's So Great About Aggregators?

Aggregators are web servers that collect and unify reports from many sources. For example, many of us are familiar with news aggregators. In the radio world, aggregators collect reports of signals, like the vehicle position reports we relayed to APRS-IS in Try This: Track Vehicles with APRS, on page 72; these reports can then be viewed on sites like APRS.fi.[1]

Flight Data Aggregators Are a Special Case

OpenWebRX+ doesn't currently support feeding aircraft flight data to aggregators, probably because they are mainly commercial sites, and they tightly control their data feeds to ensure the integrity of the data and their ability to manage it in a professional manner.

But these companies often seek data feeds from users around the world to improve their network coverage. Some provide the equipment to let you do so. If you are interested in another side project, check out these sites:

Flightradar24,[2]   FlightAware,[3]   ADS-B Exchange,[4]   and ADSB Support[5]

---

1. https://aprs.fi
5. https://www.adsbsupport.com/join-our-adsb-network/
4. https://www.adsbexchange.com/ways-to-join-the-exchange/
3. https://www.flightaware.com/adsb/flightfeeder/
2. https://www.flightradar24.com/add-coverage

No Need to Beat Them, Just Join Them

So much for flight tracking aggregators ... what about sharing the other data that we decode?

Let's go to Settings > Spotting and reporting, and have a look around. This is where we can set how we share our coded data. There are four sections that we'll cover here:

- APRS-IS settings
- PSKReporter settings
- WSPRnet settings
- MQTT settings

(Lurking at the bottom, trying not to draw attention to itself, is another section, RigControl settings; this has nothing to do with the aggregators.)

Try This: Enable Background Decoding

Background decoding is an optional step here. If you don't enable it, decoding will happen only when you have OpenWebRX+ opened in your browser. You might be fine with this, or you might prefer to leave the OpenWebRX+ host running 24/7 and decoding while you sleep. It's up to you.

Go to Settings > SDR device settings > RTL-SDR, and under "Device settings," ensure that the "Enable this device" and "Keep device running at all times" are ticked. Then go down to "Additional optional settings," select "Run background services on this device" from the drop-down list, and click the Add button. Then tick the new "Run background services on this device" checkbox.

Under "Additional optional settings," consider selecting "Require magic key to switch profiles on this device," clicking the green Add button, and ensuring the new checkbox is ticked. This will stop anyone accidentally changing the band the SDR is monitoring as it decodes data.

Finally, click "Apply and save" to make the changes permanent.

Next, we want to enable background decoding. Without it, the SDR would only decode signals while we have a browser tab open and watching the decoder. So, in Settings > Background decoding, first tick the "Enable background decoding services" checkbox.

Next, we need to decide which kinds of signals we want to decode. We can see a long line of checkboxes, one for each supported decoder. While it's tempting to tick them all, keep in mind that each decoder places a load on the Pi's processor, so it's best to check only the ones you are specifically interested in. You can come back and change your selections here any time.

Remember also that the RTL-SDR can deal with only a 2.5MHz chunk of spectrum, so, for example, it could not hear NOAA satellite signals on 137MHz and VHF Amateur band APRS signals on 145MHz at the same time; there would be no point in having both decoders running simultaneously.

If we were to add more RTL-SDRs, then we could have one monitoring NOAA and the other monitoring APRS, but we're not there yet.

Try this: Create an APRS IGate

An IGate (Internet Gateway) takes received APRS transmissions and relays them to the APRS Internet Service;[6] from there, websites such as the popular APRS.fi[7] site in Finland present the received APRS data on an interactive map.

Hams Only

To run an IGate, you must hold an amateur radio license, as APRS-IS requires an amateur radio call sign for access.

You also need an APRS-IS network passcode; strangely, there's no formal procedure to apply for one. Instead, you can use an online APRS-IS Passcode Generator, give it your call sign, and you will get a passcode back.

In Settings > Background decoding, tick the Packet checkbox in the "Enabled services" list.

In Settings > Spotting and reporting, fill out the "APRS-IS settings" section:

- APRS callsign—Enter your amateur radio call sign; some operators append a substation ID such as "-1" to distinguish their IGate from their other APRS transmissions.

- APRS-IS server—The APRS Tier 2 Network page[8] lists the regional servers. Pick the one nearest you and append the "client-defined filter" port number, :14580. For example, in Australia and New Zealand, we use the Oceania server, so we would enter "aunz.aprs2.net".

- APRS-IS network password—Your network passcode, as described above

---

6. https://www.aprs-is.net/
7. https://aprs.fi/
8. https://www.aprs2.net/

- Optionally, if you have set up your OpenWebRX+ receiver location in General Settings and you are happy for it to be displayed online, tick the "Send the receiver position to the APRS-IS network" checkbox.

- APRS beacon symbol—Pick "Receive only IGate (R&)" from the drop-down list; the IGate will appear on the APRS.fi map as a black diamond with the letter R (for repeater).

- APRS beacon text—This can be any short descriptive text, for example, "OpenWebRX+ APRS IGate."

- Antenna height, Antenna gain, and Antenna direction can be left blank.

Click the "Apply and save" button.

Make sure that your RTL-SDR is switched to a profile that includes your local APRS frequency.

It may take a minute or two, but if you have allowed your receiver location to be sent to APRS-IS, then you'll see its marker on the APRS.fi map, tagged with your call sign. Click the marker and then "info" to see more information about your new IGate.

OpenWebRX+ transmits identification packets to APRS-IS once an hour or whenever you change something. Also, any APRS transmissions that your IGate hears will be relayed to the servers, so you'll be able to see the movements of passing vehicles and more.

Congratulations, you are now feeding receiver data to the world!

On APRS.fi, the tracks of moving vehicles will appear, and if you hover over dots on the track, you'll see lines to the IGates that provided the data feed; hopefully, your IGate will be among them!

Try This: Discover Signals in and out of Your Area

PSK Reporter is another website that accepts reports of digital radio transmissions and displays them on a map. The sending and receiving stations usually include their locations so PSK Reporter can build up a picture for any place on Earth, showing the flow of radio signals in and out of the area, as shown at the top of the next page.

We can feed our own signal reports to this site as well, but first, let's see how we can make a map for our location.

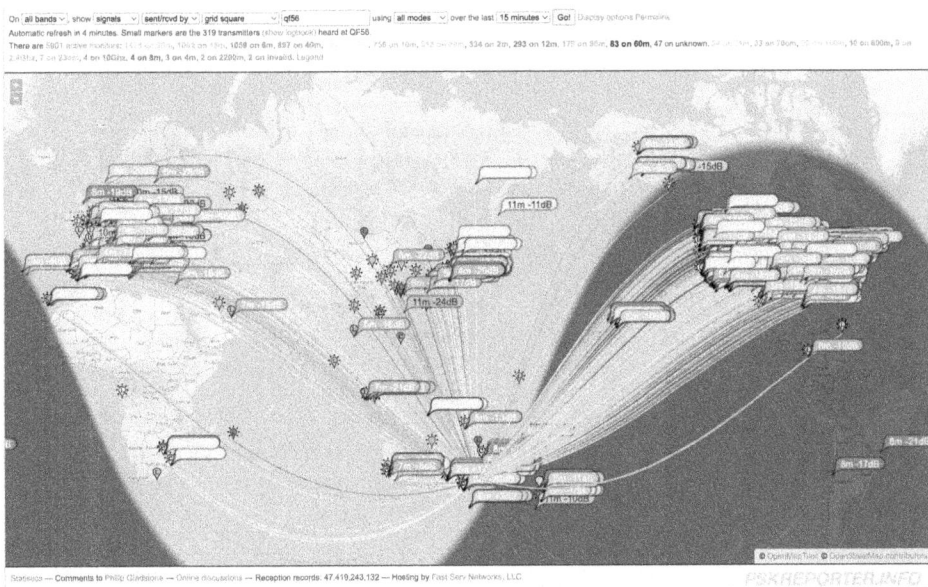

- Open a new browser tab, and go to the PSK Reporter propagation map.[9]

- Next, we need to find our Maidenhead Grid Square, which we met in About Maidenhead Locators, on page 31:

    - Click Display Options.
    - Tick the "show grid" checkbox.
    - Close the Display Options dialog box.

- Zoom the map in to your location (or any other area of interest), and note the four-character code in the center of the box. This is your Grid Square; make a note of it.

- Open Display Options again, and set the options, as shown on the next page.

- Close Display Options again.

- Set the line at the top of the screen to read, "On all bands, show signals, sent/rcvd by, grid square, (your four-character grid square) using all modes over the last 15 minutes."

- Click the Go! button. The map should look like the one we saw a few pages back, but focused on the grid square you entered.

---

9. https://pskreporter.info/pskmap.html

Display Options ✕

☑	Hide faint monitors
☑	Hide monitors if no reports
☐	Hide pink blob
☐	Hide night shadow
☑	Hide city lights
☐	Show unseen tx
☐	Show grid
☑	Show snr
☐	Monitors in frequency order
☑	Suppress bad QRG
☐	Hide statistics
☐	Hide everying but the map
☐	Hide connecting lines
☑	Show connecting lines always
☐	Hide seen times
☐	No auto pan/zoom
☐	SNR in LogBook
☐	Show time text in Black always
Mercator ⌄	Map type
[ ]	Azimuthal center locator
10	Minutes for Sparkly markers
0.65	Darkness for night shadow (0-1)
Show all ⌄	transmitters
no ⌄	timeout for worked markers
Automatic ⌄	as distance units

If you see no reports, or only a few, it might mean that there are no (or too few) active transmitters or receivers in your area providing data to create the map. Here are a few things you can try:

- Change the locator to only the first *two* characters, for example. QF instead of QF56; this shows reports over a wider area that still includes your location, so the numbers should improve.

- Wait a while. Many digital stations run without a human to press the "transmit" button, but if all yours are manually operated, then you might need some time.

- You might be unlucky and live in a part of the world where there are no active receivers at all, in which case you'll be a pioneer for your area! Head over to Chapter 6, Share Your Decoded Data, on page 77 to take some action!

If you still don't see plenty of signals, try using the locator of your nearest large town or city; these are where there will probably be more transmitters and receivers providing data to PSK Reporter.

It's also possible that the Sun has decided to unleash a flare, thereby causing a radio blackout; check the Solar Ham website[10] to confirm this. In this case, you have to wait until the ionosphere settles down. Don't worry, it will pass, but it could take a few hours.

When you are happy with the map, click Permalink (to the right of Display Options) and bookmark the page in your browser.

When you hover over any of the markers on the map, you'll see information about the sending and receiving station, the frequency, and more, all at the bottom left of the window.

It's also fun to watch how the overall pattern changes throughout the day and night. This gives us insights about how the ionosphere is changing, and which amateur radio bands are working well to make radio contact with different parts of the world. The markers are color-coded by band to make it easy to see at a glance.

Try This: Send Signal Reports to PSK Reporter

PSK Reporter accepts signal reports for a wide range of bands. Click the "all bands" drop-down box at the top left to see them all. Generally, the shortwave bands between 80 meters and 6 meters offer long-range communications, so they'll be easier to see on a world-scale map. With the others, you'll need to zoom in quite a bit.

First, though, you need to receive the signals that PSK Reporter tracks. On the Reporter, open the "all modes" drop-down list to see them. They may be sorted by popularity, which is why FT8 is at the top. Check the Signal Identification Guide to see the popular frequencies, and see if you can hear any stations. If so, you're ready to report them.

In Settings > Background decoding, tick the FT8 checkbox in the "Enabled services" list and any other modes of interest you saw on PSK Reporter.

After a while, your received signals should start appearing on the map.

---

10. https://www.solarham.com/

No Amateur Radio License? No Problem!

Unlike the APRS Internet Service, you don't need to have an amateur license to feed data to PSK Reporter. According to the web site,[11] you can make up your own identifier.

The recommended format is: PREFIX/SWL/PLACE, where PREFIX is the amateur radio prefix[12] for your area, SWL is short for "Short Wave Listener," and PLACE is the name of your city or region.

So, if you are in Boston, you might use W/SWL/BOSTON; in Sydney, Australia, perhaps VK/SWL/SYDNEY, and so on.

Try This: WSPRnet

WSPRnet is similar to PSK Reporter, though it's more focused on WSPR and a few closely related modes. You can send your received WSPR reports for inclusion on the WSPRnet map.[13]

Like APRS-IS, you need an amateur radio call sign, which you enter in the "wsprnet callsign" field, and tick the "Enable sending spots to wsprnet.org" checkbox.

In Settings > Background decoding, tick the WSPR checkbox in the "Enabled services" list and any other modes of interest you saw on PSK Reporter.

Set your RTL-SDR to a band profile that includes WSPR transmissions, and OpenWebRX+ will take care of the rest.

MQTT settings

MQTT, previously known as Message Queuing Telemetry Transport—even though no queues are involved—is a messaging protocol for publishing data from IoT (Internet of Things)[14] devices, such as environmental sensors, and for letting others subscribe to those data feeds.

OpenWebRX+ can act as the IoT device in this case, so MQTT is another mechanism to feed that to others.

---

11. https://pskreporter.info/
12. https://www.qsl.net/xe2nat/prefijos.htm
13. https://www.wsprnet.org/drupal/wsprnet/map
14. https://studyonline.unsw.edu.au/blog/what-is-iot

We won't get into the protocol here, but you can find out more on the MQTT main site[15] or check out one of the MQTT tutorials[16] online.

Up Next: Share Your Receiver with the World

Sharing your decoded data is one thing, but we haven't yet talked about sharing your entire receiver with others online. Whether you decide to do this is up to you, of course, but if you do, you'll have plenty of company. Let's see how ... it's easier than you'd expect!

---

15. https://mqtt.org/
16. https://www.hivemq.com/blog/how-to-get-started-with-mqtt/

CHAPTER 7

Take Your WebSDR Public

One of the great features of OpenWebRX+ is that we can open up the receiver so that anyone in the world can listen to it just as we can listen to all the other web-connected receivers.

You don't have to do this, of course, but if you'd like to join this particular radio community, it's easy to do.

As Easy as 1, 2, 3 ... 4, 5

Opening up your OpenWebRX+ to listeners online involves five easy steps, which we'll cover in the following exercises.

First, we'll lock our OpenWebRX+ host's IP address on our local network.

Then, we'll tell our router to take any external requests it receives on port 8073 and forward them to our host.

After that, we'll register a domain name on Duck DNS[1] that points to our router's external IP address.

We're almost done; we'll set up a script to notify Duck DNS of any changes to our external IP address.

Finally, we'll get our receiver listed on the Receiverbook directory, to make it easy for listeners to find us.

A Fixed Abode for Your OpenWebRX+ Host

Any time we reboot our router or the OpenWebRX+ host, there is the possibility that its IP address on our local network will change. Normally, we don't

---

1.   https://www.duckdns.org

care about the IP address, because we access the host by its name—for example, ssh vk2sky@openwebrxplus-pi.local.

We can override this behavior by reserving a permanent IP address on our local network for our host machine.

Try This: Reserve a Local IP Address

In this exercise, we'll lock our OpenWebRX+ host to a fixed local IP address. This will be important in the next exercise when we forward port 8073 requests on our public address to the OpenWebRX+ host.

Mastering Your Wireless Router: The Ultimate Guide To Setting Up DHCP Reservations![2] is a good, quick introduction to reserving an IP address.

But each manufacturer's router does things slightly differently, so do a search for your router model and "DHCP reserve address"; this should bring up more specific instructions for finding the Address Reservation section on your particular router.

The router maps the host's MAC Address, which is determined by the host hardware, to its internal IP address. If you ever change the host hardware, the new one will have its own MAC Address, so you'll have to modify the Address reservation you created for the old host.

As an example, on a TP-Link router, clicking the Advanced tab, then Network on the side menu, then LAN Settings, shows information about the devices connected to the router. The Client List shows all the devices with their MAC address and their current IP address on the network:

Client List

Total Clients: 11   ↻ Refresh

ID	Client Name	MAC Address	Assigned IP	Leased Time
1	wireless-gateway	4C-AC-x-x-xx/x	192.168.1.100	23:49:20
2	Mike-workbook-3	2x-xx-xx-xx-xx-x	192.168.1.101	23:49:20
3	Canon-xx-xx-co-x-xx-b	98-77-xx-xx-x-xx	192.168.1.105	23:49:23
4	Roku	E-x-EA-Cx-xx-xx-x-OP	192.168.1.102	23:49:24
5	OpenWebRX+ Pi 4	DC-A6-32-A4-E1-CF	192.168.1.103	23:49:29

‹ 1 2 3 ›

---

2. https://www.youtube.com/watch?v=dCAsHdRBrag

We can see the Raspberry Pi 4 running OpenWebRX+ with the IP address 192.168.1.103 that was automatically assigned by the router and is subject to change. We'll give the Pi a fixed IP address of 192.168.1.73, though we can use any valid IP address as long as it's not used by another device on our network.

Below the Client List on the TP-Link router is the Address Reservation section. To reserve the IP address:

- Click the Add button in the Address Reservation section.
- Copy the Pi's MAC address from the Client List to the Address Reservation section.
- Enter the desired IP address, 192.168.1.73.
- Click Save.

When we reboot the Pi and refresh the router's Client List page, we can see that the Pi has the new IP address, and it's marked as Permanent.

In the Address Reservation section, we can also see the Pi's reservation details, which we can edit or delete later if required:

Client List

Total Clients: 11				⟳ Refresh
ID	Client Name	MAC Address	Assigned IP	Leased Time
1	~~~~~~	~~~~~~	192.168.1.100	23:46:01
2	~~~~~~	~~~~~~	192.168.1.101	23:46:01
3	~~~~~~	~~~~~~	192.168.1.105	23:46:04
4	~~~~~~	~~~~~~	192.168.1.102	23:46:05
5	OpenWebRX+ Pi 4	DC-A6-32-A4-E1-CF	192.168.1.73	Permanent

◁ 1 2 3 ▷

Address Reservation

⊕ Add ⊖ Delete

☐	MAC Address	Reserved IP Address	Group	Status	Modify
☐	DC:A6:32:A4:E1:CF	192.168.1.73	Default	♡	✎ 🗑

Port Request: Pass It On!

Our host now has a fixed IP address on our local network, 192.168.1.73, but it's not yet accessible to the outside world. As proof, let's find our public IP address and try to access port 8073:

- In your web browser, go to "https://whatismyipaddress.com/"; this will display your public IPv4 address, something like 123.45.67.89.

- Take that IP address, add ":8073" to the end, and try to open that URL. The web browser will show an error such as ERR_CONNECTION_ REFUSED.

So, there is no link between (your public IP address):8073 and 192.168.1.73:8073. Let's make that work now, by using port forwarding.[3]

Try This: Forward Public Port 8073 Requests

If you're unfamiliar with the concept of port forwarding, the video Beginners Guide to Port Forwarding[4] will convey the basics.

Again, different router manufacturers do this slightly differently; search online for your router model and terms like "NAT" (network address translation), "port forwarding," or "virtual servers."

Again, using a TP-Link router as an example, How to Set Up Port Forwarding on a TP-Link router[5] explains the basic concepts. The video example uses port 25, but we will use the OpenWebRX+ port, 8073, instead. Here's the procedure:

- Log in to the router.
- Click the Advanced tab.
- Select NAT Forwarding from the menu on the left.
- Select Virtual Servers.
- Click Add.
- You can leave the Interface Name as it is, but set the other fields as shown here:

---

3. https://portforward.com/
4. https://www.youtube.com/watch?v=jfSLxs40sIw
5. https://www.youtube.com/watch?v=2tIUtsOfyFk

Virtual Servers

	ID	Service Type	External Port	Internal IP	Internal Port	Protocol	Status	Modify
--	--	--	--	--	--	--	--	--

Note: Virtual Server can be configured only when there is an available interface. If the external port is already used for Remote Management or CWMP, Virtual Server will not take effect.

Interface Name:	ewan_pppoe	
Service Type:	HTTP	Scan
External Port:	8073	(XX-XX or XX)
Internal IP:	192 . 168 . 1 . 73	
Internal Port:	8073	(XX or Blank, 1-65535)
Protocol:	TCP	

Enable This Entry

Cancel   Save

Finally, click the Save button.

Now, go back and try (your public IP address):8073; this time, your OpenWebRX+ receiver should start. Congratulations, you've gone public! Should you ever change your mind, all you have to do is disable or delete port forwarding on the TP-Link router. Clicking the Status icon for the item you just created will disable it, and another click will enable it again. Or, you can use Delete to remove the link entirely.

What's in a (Domain) Name?

So, now you can access OpenWebRX+ using your public IP address and port 8073, but as you saw when accessing a device on our local network, a name is much easier to remember. Let's give your public IP address a domain name.

One easy way to register a domain name is to use a Dynamic DNS (Domain Name Server) service, which will give us a URL like "http://my-openwebrx-receiver.some-dynamic-dns-provider.com." Technically, it's a subdomain, since the domain name proper is that of the DNS service.

Try This: Register a Domain Name

A popular (and free) Dynamic DNS provider is Duck DNS,[6] and there are online tutorials for using it, such as How to Setup Duck DNS with Raspberry Pi (EASY).[7]

We'll go through the basics here:

- Go to the Duck DNS website and sign in with Persona, Twitter (X), GitHub, or Google.
- Complete the reCaptcha.[8]
- Fill in the sub-domain field with the name you'd like your receiver to be known by, for example, "vk2sky-openwebrx" (but that name is taken, by me, so choose your own).
- Click the green "add domain" button.
- Now, the subdomain vk2sky-openwebrx.duckdns.org has been created!

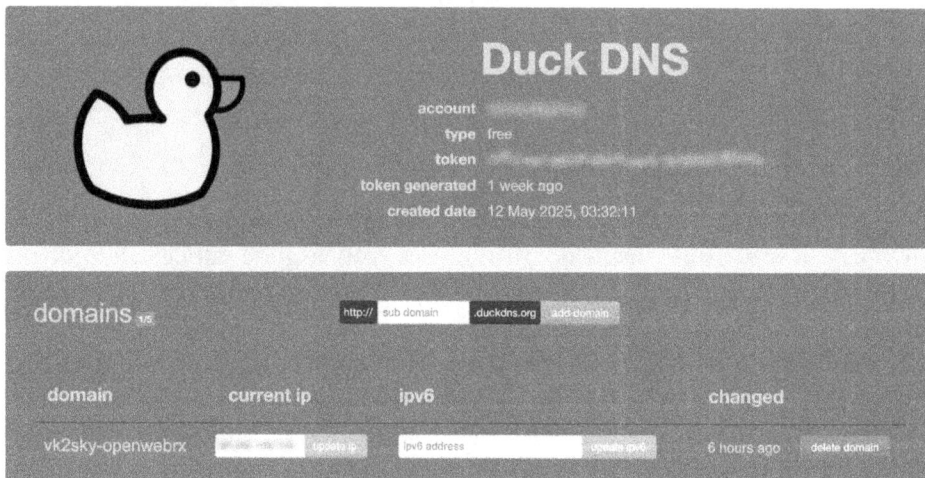

Here are a few other things to notice on the updated screen:

- Near the top, there's a unique token associated with your Duck DNS account; there's no need to memorize or copy it, but we will see it again shortly.

- You'll see a list of domains that were added (so far, only one). Each entry has:

---

6.   https://www.duckdns.org/
7.   https://www.youtube.com/watch?v=z092FGtP6ME
8.   https://developers.google.com/recaptcha

 – a domain (for example, "vk2sky-openwebrx");

 – a "current ip" that we associated with the domain name; that is, the public IPv4 address for our router, and a yellow "update ip" button;

 – an ipv6 address and yellow "update ipv6" button, which we can ignore;

 – how long ago the domain entry was last changed; and

 – a red "delete domain" button.

You can change the IPv4 address at any time or delete the domain if you no longer need it. You can also add further domains if you wish, in the same way that you added this one.

While we're noting things, there's a "donate" button at the bottom left of the window where we can help support this free service.

Let's check if the new domain name works. In your browser, go to "http://(your-domain-name).duckdns.org:8073" and your OpenWebRX+ receiver should appear!

For more information on this topic, see Raspberry Pi DDNS Setup Tutorial! (Duck DNS).[9]

Handle Public IP Address Changes

So, we've connected a friendly domain name to our public IP address, which is linked to the host's internal IP address by port forwarding. Great, but as we saw a little while ago, our public IP address could change, so we have one last problem to solve.

When we notice that our domain name is no longer reaching our router, we could find out our new public IP address, go to the Duck DNS website, and use the "update ip" button, but there's an easier way.

Try This: Keep the Duck in the Loop

Instead, we'll get our OpenWebRX+ host to regularly inform Duck DNS of our current public IP address, so that it knows when the address changes.

At the very top of the Duck DNS page is a menu: click Install. Below the duck image, we can now see an Operating Systems section. We're using a Raspberry Pi, so click the "pi" button.

---

9. https://www.youtube.com/watch?v=s-66gmlHoyE

Below that, we now see "first step - choose a domain," and underneath that, "http://," a "choose domain" drop-down list, and ".duckdns.org." From the drop-down list, choose the domain name you entered earlier, for example, "vk2sky-openwebrx."

Then, a series of terminal instructions appears. SSH into your OpenWebRX+ host and copy/paste those instructions with a few minor changes.

```
➜ ~ ssh vk2sky@openwebrxplus-pi.local
```

The Duck DNS instructions suggest creating a directory, duckdns, under your home directory, but you'll instead create it in the /opt directory (security tip courtesy of VK2BTQ). In that directory, you'll create your script file, duck.sh. Because /opt is not in your personal user space, you need to use the sudo command prefix:

```
$ sudo mkdir /opt/duckdns
$ cd /opt/duckdns
$ sudo nano duck.sh
```

Then, copy the next block of text from the Duck DNS install page, and paste it into your terminal window. It's a long line, but don't try to reformat it; copy and paste as-is. The text looks like the following, but the "domains=..." and "token=..." parts will be unique for your installation:

```
echo url="https://www.duckdns.org/update?domains=...&token=...&ip=" | curl -k -o...
```

Make one minor edit to the end of that line: change ~/duckdns/duck.log to /opt/duckdns/duck.log, so that the log file stays in the same directory as the script.

Save the file and exit the editor.

You also need to make the shell script executable, using the chmod command:

```
$ sudo chmod 700 duck.sh
```

Next, test the script:

```
$ sudo ./duck.sh
```

The output will look something like this:

```
  % Total    % Received % Xferd  Average Speed   Time    Time     Time  Current
                                 Dload  Upload   Total   Spent    Left  Speed
100     2    0     2    0     0      1       0 --:--:--  0:00:01 --:--:--     1
```

To see if the script worked, check the resulting log file:

```
$ cat duck.log
```

You should see "OK," followed by the prompt for the next terminal command. If you see "KO" instead of "OK," then something has gone wrong, and you should go back and check that the line you added to duck.sh looks exactly as described.

Next, tell the OpenWebRX+ host to execute this script every five minutes. To do this, you need to edit the host's table of cron jobs, using the crontab command:

```
$ crontab -e
```

Chances are you haven't run the crontab command on the host before, so you'll probably be asked to pick an editor:

```
no crontab for (your user name here) - using an empty one

Select an editor.  To change later, run 'select-editor'.
  1. /bin/nano        <---- easiest
  2. /usr/bin/vim.basic
  3. /usr/bin/vim.tiny
  4. /bin/ed

Choose 1-4 [1]:
```

Press Enter to use nano, or pick one of the other editors.

Copy the following line and paste it at the end of the file. Remember that while the Duck DNS instructions specify ~/duckdns, your duck.sh is located in /opt/duckdns:

```
*/5 * * * * /opt/duckdns/duck.sh >/dev/null 2>&1
```

Again, save and exit the editor.

Here's a quick translation of the cryptic command here: "every five minutes, run the script /opt/duckdns/duck.sh, and disregard anything it prints out."

If you'd like to explore crontab expressions further, crontab guru[10] is a great place to start.

Next, start the cron service:

```
$ sudo service cron start
```

Those five-minute updates should now be occurring. The cron service should also automatically start if you reboot the host, so you don't have to worry about running it manually.

---

10. https://crontab.guru/

To summarize:

- Your OpenWebRX+ host resides at a fixed IP address on your internal network.
- You have configured your router to take any request appearing on port 8073 of your public IP address and forward it to the OpenWebRX+ host on your internal network.
- You used Duck DNS to register a domain name for your receiver, so that you can access the receiver from the Internet using that name instead of the numeric IP address.
- Every five minutes, your host informs Duck DNS of your current public IP address.

Get Your Receiver on the List (and Map)

So, now you can share your receiver's URL with anyone you wish. Remember to include the port number, :8073, at the end.

If you only want to share your receiver with a "chosen few," you can stop here. But, if you'd like to share it with the world at large, there's one more step: getting listed and mapped on an OpenWebRX online directory.

Try This: Register Your Receiver on Receiverbook

A few simple steps will get your receiver listed online:

- In your browser, go to Receiverbook.[11]

- Click Login/Register and "Sign up" to create an account.

- On the top menu, click "Report a new receiver."

- In the URL field, enter the receiver's domain name, which you created in the previous exercise, and the port number: "http://(your domain).duckdns.org:8073."

- Click the "Submit new receiver..." button.

The claim status page that appears assigns you a unique receiver key that looks like "receiverbook-" followed by a string of hexadecimal digits, shown on the opposite page.

Copy the receiver key. Then, in OpenWebRX+, go to Settings > General, and scroll down to the "Receiver listings" section. Paste the key into the "Receiver keys" box, and click "Apply and save" at the bottom.

---

11. https://www.receiverbook.de/

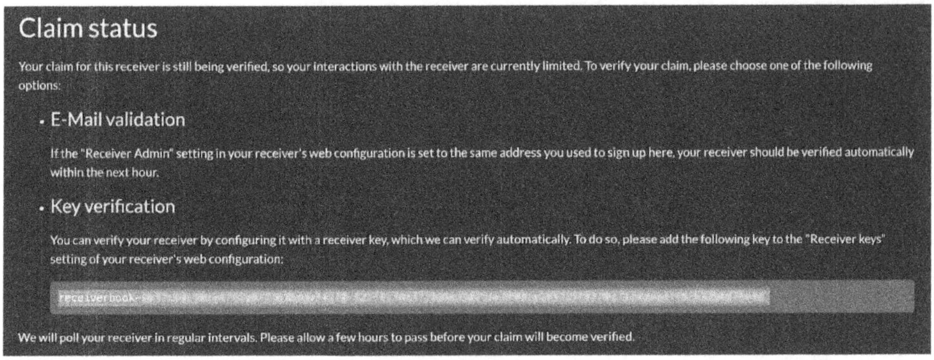

Claim status

Your claim for this receiver is still being verified, so your interactions with the receiver are currently limited. To verify your claim, please choose one of the following options:

- **E-Mail validation**

  If the "Receiver Admin" setting in your receiver's web configuration is set to the same address you used to sign up here, your receiver should be verified automatically within the next hour.

- **Key verification**

  You can verify your receiver by configuring it with a receiver key, which we can verify automatically. To do so, please add the following key to the "Receiver keys" setting of your receiver's web configuration:

  receiverbook-xx ...

  We will poll your receiver in regular intervals. Please allow a few hours to pass before your claim will become verified.

Then, we wait ... it can take up to a few hours for Receiverbook to complete the registration, but pretty soon, your receiver will appear on the Receiverbook map.[12]

If you click Online Receivers > List view, scroll to the bottom of the page, and click the ">|" button at the bottom right. You should see your receiver's details listed there. Mission accomplished!

_____

12. https://www.receiverbook.de/map

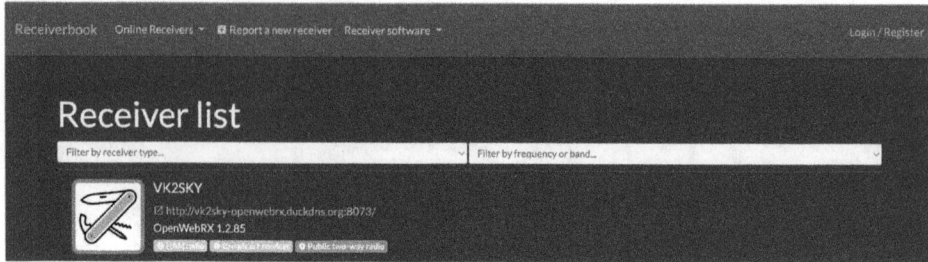

Up Next: Extend OpenWebRX+ (and Yourself)

Congratulations, you now have a publicly listed, web-connected radio receiver! You could stop here, but being public presents a few interesting challenges and new areas to discover and explore. Take a break if you like, and when you're ready, jump back in!

CHAPTER 8

Go Above and Beyond

We've come a long way in a short time with our web-connected receiver, but there is still room for improvement, and OpenWebRX+ still has a few tricks up its sleeve.

Two (Radio) Heads Are Better Than One

The RTL-SDR is a great entry-level receiver, but it can handle only 2.5MHz of spectrum at any one time. This can become a significant limitation when online visitors start using our receiver. Let's look at how we can deal with that.

For example, what if one user wants to listen to our FM broadcast station on 92.9MHz, and another wants to listen to amateur radio activity around 145MHz? The RTL-SDR can tune both of these frequencies, but not at the same time because they are more than 2.5MHz apart.

One solution is to connect another RTL-SDR to our host, or even add multiple RTL-SDRs. Another is to upgrade to a more expensive receiver that can cover a larger frequency range at once; we can add multiples of these receivers, too. Let's try both approaches.

Try This: Add Another RTL-SDR

Spoiler: it's not quite as simple as plugging in another dongle. For a start, getting two or more to fit directly onto the Pi is impossible, so we'll probably need to buy an extension cable or USB hub. Another factor is that we need two antennas, but if the two RTL-SDRs were going to operate on vastly different frequencies, we'd probably have to do that anyway. If the two receivers will operate on similar frequencies, they can share one antenna using a device that goes by various names such as "antenna combiner," "RF power (or signal) splitter," "RF signal distributor," "combiner/splitter," and so on. For a cheap experiment, try an SMA female to twin SMA male adaptor cable.

Next, when we have two or more RTL-SDRs connected to our Raspberry Pi, OpenWebRX+ needs a little help to tell them apart. Go to the first RTL-SDR's device settings page, scroll down to the "Recent device log messages" section, and look for the words, "Found 2 device(s):"

```
- owrx.source.rtlsdr - WARNING - source has not shut down normally within 10 seconds, sending SIGKILL
- owrx.source.rtlsdr - INFO - Started sdr source: rtl_connector -e 0 -s 2500000 -g 29 -d 0 -f 438800000 -p 5039
- owrx.source.rtlsdr - INFO - STDOUT: setting up control socket...
- owrx.source.rtlsdr - INFO - STDOUT: control socket started on 53489
- owrx.source.rtlsdr - INFO - STDOUT: socket setup complete, waiting for connections
- owrx.source.rtlsdr - INFO - STDOUT: Found 2 device(s):
- owrx.source.rtlsdr - INFO - STDOUT:   0:  RTLSDRBlog, Blog V4, SN: 00000001
- owrx.source.rtlsdr - INFO - STDOUT:   1:  Realtek, RTL2838UHIDIR, SN: 00000001
```

In the previous example, "0: RTLSDRBlog, Blog V4..." is an RTL-SDR v4 (no surprise there); the other is an older v3 that identifies more cryptically as "1: Realtek, RTL2838UHIDIR..." If they were two identical models, they would have the same description, but each would still have a unique "Device identifier" number: 0, 1, and so forth.

Next, scroll up a bit to "Additional optional settings," and open the drop-down list. Find the entry, "Device identifier," select it, and click the Add button, so it moves up to be with the other settings. Enter the number 0 in the box, and click "Apply and save."

Next, we'll create a device profile for the second RTL-SDR and configure it similarly to the first one. Click Settings > SDR device settings, then click the "Add new device..." button at the bottom.

The settings will be similar to the first RTL-SDR, but give it a unique device name such as "RTL-SDR 2." Then, pick the "Device type," "RTL-SDR device," and click "Apply and save."

The screen refreshes and reveals the new device settings. As with the first RTL-SDR, pick "Device identifier" from the "Additional optional settings," but this time set it to 1.

The other settings can be the same as the first RTL-SDR.

We can then set up whatever band profiles we like, as we did for the first RTL-SDR in Try This: Create a Band Profile, on page 22.

Back on the main OpenWebRX+ receiver page, if you open the band/profile drop-down list, you should be able to see both RTL-SDRs.

Think Bigger SDRs

Another approach is to buy a more powerful SDR that can grab more bandwidth. The OpenWebRX+ Hardware Guide[1] has details of some of the more popular SDR receivers, including the RTL-SDR, and instructions for making them work with OpenWebRX+. Depending on the bands you want to explore and how much bandwidth you need to grab, one of those should be right for you.

The "non-plus" version of OpenWebRX lists other supported SDR receivers[2] that may be worth exploring, too. The device-specific notes section of that page will take you to the details for each receiver.

---

1. https://fms.komkon.org/OWRX/#HardwareGuide
2. https://github.com/jketterl/openwebrx/wiki/Supported-Hardware#sdr-devices

Another consideration here is that if you start grabbing and processing more spectrum, you may need to upgrade your host to a more powerful model.

Add More Demodulators

Way back in Install the Software, on page 9, you might have noticed this comment on the OpenWebRX+ releases page:

> Please note that these images come without the software support for digital modes (DMR, NXDN, etc), since the software decoder for these modes (mbelib) has questionable origins. In order to enable digital modes in these images, ssh into the user account you created while installing the image and type: [...] sudo install-softmbe.sh

The phrase "has questionable origins" roughly translates thus: nobody seems to know for sure whether the intellectual property used to create it was legally obtained. You can read a discussion about the matter on RadioReference[3] and make your own decision.

Try This (Maybe): Add Extra Digital Modes

If you're comfortable with the legal status of MBELIB, it's easy to install: SSH into your OpenWebRX+ host, and enter the command:

```
$ sudo install-softmbe.sh
```

You will see that your receiver panel now sports a few new demodulators:

- DMR (Digital Mobile Radio)[4]
- D-STAR (Digital Smart Technologies for Amateur Radio)[5]
- NXDN (Next Generation Digital Narrowband)[6]
- YSF (Yaesu System Fusion)[7]

Explore the ISM Bands

ISM is short for the Industrial, Scientific, and Medical bands that are used for applications in these and other fields. Typically, they are electronic devices transmitting control signals, telemetry, and other short-range data.

---

3. https://forums.radioreference.com/threads/mbelib-imbe-ambe-patent-questions.437024/
4. https://www.sigidwiki.com/wiki/Digital_Mobile_Radio_(DMR)
5. https://www.sigidwiki.com/wiki/D-STAR
6. https://www.sigidwiki.com/wiki/NXDN
7. https://www.sigidwiki.com/wiki/Yaesu_System_Fusion

The actual ISM frequencies vary around the world, so you'll need to do a little research to find out where they are. ISM Bands Around the World[8] is a good place to start.

Try This: Monitor Your Home Weather Station Sensors

Many of us have home weather stations that include a wireless outdoor temperature and humidity sensor. We can use OpenWebRX+ to monitor these sensors.

Depending on where you are, the ISM bands are in different parts of the spectrum.

The International Telecommunication Union's *Radio Regulations Articles, Edition of 2020*[9] document lists these:

Band	Center Frequency	Notes
6.765–6.795MHz	6.780MHz	
13.553–13.567MHz	13.560MHz	
26.957–27.283MHz	27.120MHz	
40.66–40.70MHz	40.68MHz	
433.05–434.79MHz	433.92MHz	Region 1 (Europe) with exceptions
902–928MHz	915MHz	Region 2 (Americas)
2.400–2.500GHz	2.450GHz	
5.725–5.875GHz	5.800GHz	
24–24.25GHz	24.125GHz	
61–61.5GHz	61.25GHz	
122–123GHz	122.5GHz	
244–246GHz	245GHz	

Ideally, your sensor will be marked with its operating frequency, or at least its band, but you may still need to do some research and experimenting to find out exactly where it is transmitting.

In Australia, one of the ISM bands covers 915–928MHz, several times as much as an RTL-SDR can grab at once. My weather sensor is marked "917 MHz," which narrows it down a lot, but it also has a seven-way channel selector switch.

8.   https://resources.altium.com/p/ism-bands-around-world

9.   https://search.itu.int/history/HistoryDigitalCollectionDocLibrary/1.44.48.en.101.pdf

First, you'll need to add a new band profile for that segment. Again, refer back to Try This: Create a Band Profile, on page 22 for a refresher. Set your profile name to something like "917MHz ISM," set your center and initial frequencies, and pick "ISM" for initial modulation. In Australia, 917MHz band ISM channels are spaced 200kHz apart, so a tuning step of 50kHz (the maximum allowed) is appropriate.

My weather station has two units: an outdoor wind speed/wind direction/rain/temperature/humidity sensor on the roof and an indoor temperature/humidity sensor. After a bit of experimenting, it turns out that both transmit on 917.0MHz.

The transmissions appear as short bursts on the waterfall, and we can see the two ISM devices, each with its own unique ID.

Incidentally, looking at the decoded data, although it's still transmitting data, "battery_ok … 0" suggests that the indoor sensor's battery could do with replacement. Hooray for telemetry!

Even if you don't have domestic ISM sensors, there is more to explore here. On YouTube, See Meshtastic Signals Using Openwebrx+ in Ubuntu VM on

Windows 10^{10} includes a segment (around the 31-minute mark) about decoding automotive telemetry on the ISM segment shared with the amateur 70cm band in Germany.

OpenWebRX+ has many more digital decoders to experiment with and, as usual, the Signal Identification Guide will guide you on your journey.

Ham It Up

Finally, here's another one for the radio hams. OpenWebRX+ version 1.2.70 added support for Hamlib,[11] the Ham Radio Control Library. If you are a licensed amateur radio operator, and your transceiver has a CAT (Computer Aided Tuning) interface, you can control your transceiver with OpenWebRX+.

OpenWebRX+ also supports the flrig[12] radio control library. If you wish to use either Hamlib or flrig, you need to install the software manually.

Out of the box, OpenWebRX+ also supports rigctl, which is part of the Hamlib library. That's what we'll use here.

Try This: Remote Control Your Ham Receiver

For this example, we'll use the popular ICOM IC-705; it can connect to an existing Wi-Fi network or even set up its own Wi-Fi hotspot, but for simplicity, we're going to use a USB cable.

First, let's see what serial devices are connected to our host.

```
$ ls /dev/serial/by-id -l
total 0
lrwxrwxrwx 1 [...] usb-Icom_Inc._IC-705_IC-705_12010505-if00 -> ../../ttyACM1
lrwxrwxrwx 1 [...] usb-Icom_Inc._IC-705_IC-705_12010505-if02 -> ../../ttyACM2
lrwxrwxrwx 1 [...] usb-u-blox_AG[...]_GPS_GNSS_Receiver-if00 -> ../../ttyACM0
```

We can see two virtual communication ports exposed by the IC-705—one each associated with devices ttyACM1 and ttyACM2, the radio's control port and the internal sound card. We can also see the GPS dongle associated with ttyACM0.

The port assignments can change, depending on the order in which devices were plugged in. In the previous example, the GPS was plugged in first, so it was assigned ttyACM0, while the IC-705 control and audio port ("…-if00") got ttyACM1, and the IC-705's internal GPS ("…-if02") was assigned ttyACM2.

---

10. https://youtu.be/IDbKV4-1Wrl?t=1888

11. https://hamlib.github.io/

12. https://www.w1hkj.org/

We can refer to these ports by their "ttyACMx" name, but if we unplug the radio and GPS, and reconnect them later in a different order, those device names can change.

To avoid confusion, we'll refer to the radio control port by its more verbose name, /dev/serial/by-id/usb-Icom_Inc._IC-705_IC-705_12010505-if00.

We need to tell rigctl which model radio we have connected. If we type the command,

```
$ rigctl -l
```

... we'll see a list of all the radio types rigctl knows about. To save manually searching the list, we can use the Linux grep command to show only the lines of interest. Because we're using an ICOM IC-705, let's find only the lines that contain "705":

```
$ rigctl -l | grep 705
 3085  Icom  IC-705  20230109.8  Stable  RIG_MODEL_IC705
```

The number at the start tells us that rigctl knows the IC-705 as model number 3085. We could also search by manufacturer, to see all the supported ICOM radios:

```
$ rigctl -l | grep -i icom
 3002  Icom  IC-1275  20230109.0  Stable  RIG_MODEL_IC1275
 3003  Icom  IC-271   20230109.0  Alpha   RIG_MODEL_IC271
 3004  Icom  IC-275   20230109.0  Stable  RIG_MODEL_IC275
 :
 etc
```

The -i option means "ignore whether letters are upper or lower case when matching," so "icom" will match "Icom."

We can now get rigctl to send the IC-705 commands. It's a lot to type, but it should give us instant feedback about whether the connection is working:

```
$ rigctl -m 3085 -r /dev/serial/by-id/usb-Icom_Inc._IC-705\[...]-if00 -c 164 f
```

To explain:

- -m 3085 tells rigctl that we should be connected to a model 3085 radio, an Icom IC-705.

- -r /dev/serial/by-id\[...] specifies the radio's device id.

- -c 164 selects ICOM CI-V address 164 (0xA4 or A4h in decimal). This option applies only to ICOM radios.

- f tells rigctl to read back the frequency to which the radio is currently tuned (note that there is no hyphen in front of the f, as it is a rigctl command, not an option).

rigctl should respond with the current frequency to which the radio is tuned, in Hertz. For example, if the radio is tuned to 7.085MHz, we'll see 7085000.

To tune the radio to a different frequency, say, 14123kHz, we can repeat the previous line, replacing the f at the end with F 14123:

```
$ rigctl -m 3085 -r /dev/serial/by-id/usb-Icom_Inc._IC-705\[...]-if00 -c 164 F 14123
```

The radio should respond by changing frequency.

Note the upper case "F" this time. Generally, lowercase letter commands read information *from* the radio, and uppercase letter commands send information *to* the radio.

To see all the things rigctl can do, take a look at the rigctl manual online.[13] There are a lot of commands, and they are worth exploring later.

For now, we'll let OpenWebRX+ handle the details for us. Go to Settings > Spotting and reporting > RigControl settings, and make these changes:

- Tick the "Enable sending changes to a standalone transceiver" checkbox.

- Transceiver model—Select your radio from the drop-down list.

- Transceiver CAT device—/dev/serial/by-id/usb-Icom_...

- Transceiver CI-V address—ICOM brand transceivers also use what the company refers to as CI-V (Control Interface, version 5.)[14] Your ICOM transceiver manual will tell you the correct CI-V address for your model, and you can download the *CI-V Reference Guide* for your radio from ICOM's support website.[15] The CI-V address is usually expressed as a hexadecimal code, for example, 0xA4 or A4h for the IC-705, but OpenWebRX+ uses the decimal equivalent[16] in this field, in this case 164. For non-ICOM radios, leave this field blank.

Finally, click "Apply and save."

---

13. https://hamlib.sourceforge.net/html/rigctl.1.html
14. https://www.icomeurope.com/wp-content/uploads/2020/08/IC-705_ENG_CI-V_1_20200721.pdf
15. https://www.icomjapan.com/support/manual/
16. https://www.binaryhexconverter.com/hex-to-decimal-converter

Back on the receiver tab, when you change frequency with OpenWebRX+, your transceiver frequency display will follow along. How you use this feature is up to you.

Go to the Source

As with all Open Source projects, we can take a look under the hood to see how OpenWebRX+ really works. The project repository[17] is hosted on GitHub, so we can download it, modify and improve it, add new features, and share them with others. The code is mainly Python, with a fair chunk of JavaScript, and I encourage everyone to help make OpenWebRX+ even better.

A Plug for Plug-ins

Thanks to Stanislav Lechev, LZ2SLL, there is also a collection of JavaScript plug-ins[18] to add even more features and enhancements. See OpenWebRX forum on Google Groups[19] for more information on how to use it and even contribute your own plug-ins.

But Wait, There's More …

There's more I could have written, and in fact did write, but there wasn't room to include it here.

But fear not, that material has not "gone walkabout" as we say Down Under; it's in the companion GitHub repository[20] that I mentioned at the start. You'll find bonus materials such as articles, tutorials, application notes, and more.

And so, as we reach the end of this book, your OpenWebRX+ journey is really beginning. Go exploring!

---

17. https://github.com/luarvique/openwebrx
18. https://0xaf.github.io/openwebrxplus-plugins/
19. https://groups.io/g/openwebrx/topic/javascript_plugins_in/103251000
20. https://vk2sky.github.io/

Crack Open the Secure SHell (SSH)

If, like me, you end up having a lot of Raspberry Pis or other computers sitting around doing various jobs, you don't want each one to have its own monitor and keyboard, adding needless clutter. As mentioned in Your OpenWebRX+ Server, on page 7, you can run your OpenWebRX+ server headless and access it remotely from your local machine's terminal using SSH. As an added benefit, the Pi doesn't need to be nearby; it can be anywhere in the world.

Which SSH Client?

The Secure SHell, or SSH, utility can give us access to the command line of a remote computer, assuming we have a valid user name and login on that computer. Most computers support SSH already, but if not, you can install an SSH client.

Native SSH

If your local machine is based on Linux or Unix (such as a Mac), you already have an SSH client. Open your terminal application to use it.

Windows and PuTTY

Recent versions of Windows also have an SSH client. On the taskbar, enter "ssh" into the search box to locate it. Alternatively, Windows PowerShell supports SSH.

Older versions of Windows do not include an SSH client; in this case, download and install PuTTY,[1] open it, and you're good to go.

---

1.   https://www.chiark.greenend.org.uk/~sgtatham/putty/latest.html

SSH Into the OpenWebRX+ host

You can sign into your OpenWebRX+ host from your local machine using the first user account you created.

Connecting to a Remote Computer for the First Time

The first time you SSH into a remote computer, you can expect a warning that the computer is untrusted.

If using SSH on the command line, the warning looks like this:

```
➜  ~ ssh vk2sky@openwebrxplus-pi.local
The authenticity of host 'openwebrxplus-pi.local (fe80::6...)' can't be established.
ED25519 key fingerprint is SHA256:d5Lk...EUA.
This key is not known by any other names.
Are you sure you want to continue connecting (yes/no/[fingerprint])?
```

Because you know that the computer you are accessing is your own, you can choose to trust it. When SSH asks if you're sure that you want to connect, type the word "yes" and press Enter. The connection proceeds:

```
Warning: Permanently added 'openwebrxplus-pi.local' (ED25519) to the list
 of known hosts.
vk2sky@openwebrxplus-pi.local's password:
Linux openwebrxplus-pi 6.6.31+rpt-rpi-v8 #1 SMP PREEMPT Debian [...]

The programs included with the Debian GNU/Linux system are free software;
the exact distribution terms for each program are described in the
individual files in /usr/share/doc/*/copyright.

Debian GNU/Linux comes with ABSOLUTELY NO WARRANTY, to the extent
permitted by applicable law.
Last login: Tue Jun  4 23:50:34 2024
$
```

For Windows users, PuTTY displays a similar warning, shown over:

The displayed details about the key fingerprint are not particularly important, only the name of the computer to which you are connecting, in this case, openwebrxplus-pi. You recognize this name, so you can click Accept.

In both cases, SSH on your local machine now knows this OpenWebRX+ host and won't warn you again.

You can now control the Pi via the SSH link from your local computer. For now, close the connection.

```
$ exit
logout
Connection to openwebrxplus-pi.local closed.
```

PuTTY Security Alert   ✕

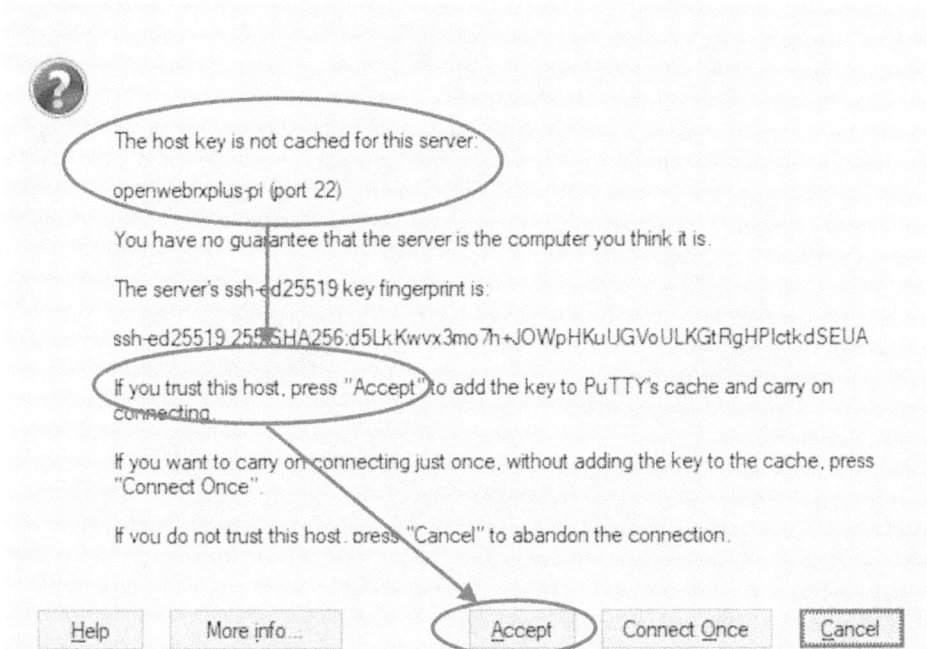

The host key is not cached for this server:

openwebrxplus-pi (port 22)

You have no guarantee that the server is the computer you think it is.

The server's ssh-ed25519 key fingerprint is:

ssh-ed25519 255 SHA256:d5LkKwvx3mo7h+JOWpHKuUGVoULKGtRgHPlctkdSEUA

If you trust this host, press "Accept" to add the key to PuTTY's cache and carry on
connecting.

If you want to carry on connecting just once, without adding the key to the cache, press
"Connect Once".

If you do not trust this host, press "Cancel" to abandon the connection.

| Help | More info... | | Accept | Connect Once | Cancel |

And that's the basics of accessing a remote computer using SSH. If you'd like to explore further, check out Learn SSH in 6 Minutes—Beginners Guide to SSH Tutorial,[2] or SSH for Beginners: The Ultimate Getting Started Guide[3] if you'd like to go into more detail.

2.   https://www.youtube.com/watch?v=v45p_kJV9i4

3.   https://www.youtube.com/watch?v=YS5Zh7KExvE

Thank you!

We hope you enjoyed this book and that you're already thinking about what you want to learn next. To help make that decision easier, we're offering you this gift.

Head on over to https://pragprog.com right now, and use the coupon code BUYANOTHER2025 to save 30% on your next ebook. Offer is void where prohibited or restricted. This offer does not apply to any edition of *The Pragmatic Programmer* ebook.

And if you'd like to share your own expertise with the world, why not propose a writing idea to us? After all, many of our best authors started off as our readers, just like you. With up to a 50% royalty, world-class editorial services, and a name you trust, there's nothing to lose. Visit https://pragprog.com/become-an-author/ today to learn more and to get started.

Thank you for your continued support.

The Pragmatic Bookshelf

SAVE 30%
with coupon code
BUYANOTHER2025
when you checkout at pragprog.com

Explore Software Defined Radio

Do you want to be able to receive satellite images using nothing but your computer, an old TV antenna, and a $20 USB stick? Now you can. At last, the technology exists to turn your computer into a super radio receiver, capable of tuning in to FM, shortwave, amateur "ham," and even satellite frequencies, around the world and above it. Listen to police, fire, and aircraft signals, both in the clear and encoded. And with the book's advanced antenna design, there's no limit to the signals you can receive.

Wolfram Donat
(78 pages) ISBN: 9781680507591. $19.95
https://pragprog.com/book/wdradio

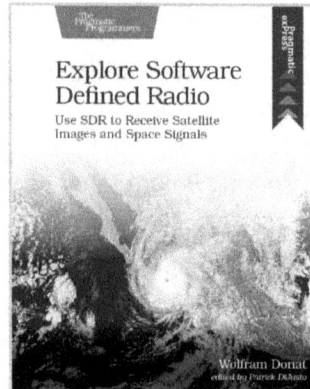

Build a Weather Station with Elixir and Nerves

The Elixir programming language has become a go-to tool for creating reliable, fault-tolerant, and robust server-side applications. Thanks to Nerves, those same exact benefits can be realized in embedded applications. This book will teach you how to structure, build, and deploy production grade Nerves applications to network-enabled devices. The weather station sensor hub project that you will be embarking upon will show you how to create a full stack IoT solution in record time. You will build everything from the embedded Nerves device to the Phoenix backend and even the Grafana time-series data visualizations.

Alexander Koutmos, Bruce A. Tate, Frank Hunleth
(90 pages) ISBN: 9781680509021. $26.95
https://pragprog.com/book/passweather

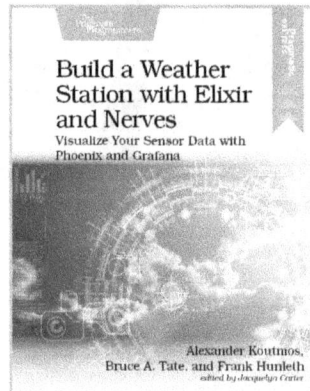

Raspberry Pi: A Quick-Start Guide (2nd edition)

The Raspberry Pi is one of the most successful open source hardware projects ever. For less than $40, you get a full-blown PC, a multimedia center, and a web server—and this book gives you everything you need to get started. You'll learn the basics, progress to controlling the Pi, and then build your own electronics projects. This new edition is revised and updated with two new chapters on adding digital and analog sensors, and creating videos and a burglar alarm with the Pi camera.

Maik Schmidt
(176 pages) ISBN: 9781937785802. $22
https://pragprog.com/book/msraspi2

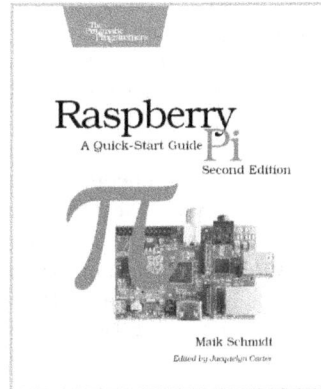

Arduino: A Quick-Start Guide, Second Edition

Arduino is an open-source platform that makes DIY electronics projects easier than ever. Gone are the days when you had to learn electronics theory and arcane programming languages before you could even get an LED to blink. Now, with this new edition of the best-selling *Arduino: A Quick-Start Guide*, readers with no electronics experience can create their first gadgets quickly. This book is up-to-date for the latest Arduino boards and for Arduino 1.x, with step-by-step instructions for building a universal remote, a motion-sensing game controller, and many other fun, useful projects.

Maik Schmidt
(322 pages) ISBN: 9781941222249. $34
https://pragprog.com/book/msard2

Automate Your Home Using Go

Take control of your home and your data with the power of the Go programming language. Build extraordinary and robust home automation solutions that rival much more expensive, closed commercial alternatives, using the same tools found in high-end enterprise computing environments. Best-selling Pragmatic Bookshelf authors Ricardo Gerardi and Mike Riley show how you can use inexpensive Raspberry Pi hardware and excellent, open source Go-based software tools like Prometheus and Grafana to create your own personal data center. Using the step-by-step examples in the book, build useful home automation projects that you can use as a blueprint for your own custom projects.

Ricardo Gerardi and Mike Riley
(160 pages) ISBN: 9798888650509. $40.95
https://pragprog.com/book/gohome

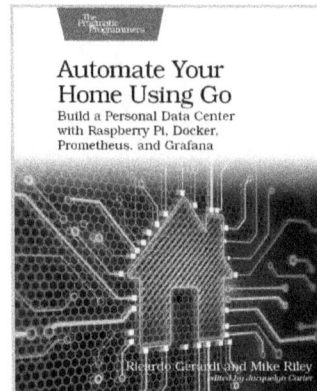

Essential 555 IC

Learn how to create functional gadgets using simple but clever circuits based on the venerable "555." These projects will give you hands-on experience with useful, basic circuits that will aid you across other projects. These inspiring designs might even lead you to develop the next big thing. The 555 Timer Oscillator Integrated Circuit chip is one of the most popular chips in the world. Through clever projects, you will gain permanent knowledge of how to use the 555 timer will carry with you for life.

Cabe Force Satalic Atwell
(104 pages) ISBN: 9781680507836. $19.95
https://pragprog.com/book/catimers

The Pragmatic Bookshelf

The Pragmatic Bookshelf features books written by professional developers for professional developers. The titles continue the well-known Pragmatic Programmer style and continue to garner awards and rave reviews. As development gets more and more difficult, the Pragmatic Programmers will be there with more titles and products to help you stay on top of your game.

Visit Us Online

This Book's Home Page
https://pragprog.com/book/rmwebrx
Source code from this book, errata, and other resources. Come give us feedback, too!

Keep Up-to-Date
https://pragprog.com
Join our announcement mailing list (low volume) or follow us on Twitter @pragprog for new titles, sales, coupons, hot tips, and more.

New and Noteworthy
https://pragprog.com/news
Check out the latest Pragmatic developments, new titles, and other offerings.

Save on the ebook

Save on the ebook versions of this title. Owning the paper version of this book entitles you to purchase the electronic versions at a terrific discount.

PDFs are great for carrying around on your laptop—they are hyperlinked, have color, and are fully searchable. Most titles are also available for the iPhone and iPod touch, Amazon Kindle, and other popular e-book readers.

Send a copy of your receipt to support@pragprog.com and we'll provide you with a discount coupon.

Contact Us

Online Orders:	*https://pragprog.com/catalog*
Customer Service:	*support@pragprog.com*
International Rights:	*translations@pragprog.com*
Academic Use:	*academic@pragprog.com*
Write for Us:	*http://write-for-us.pragprog.com*

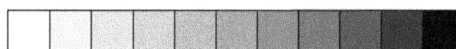

www.ingramcontent.com/pod-product-compliance
Lightning Source LLC
Chambersburg PA
CBHW081819200326

41597CB00023B/4307